Linear Algebra
for Data Science

Linear Algebra
for Data Science

Moshe Haviv

The Chinese University of Hong Kong, Shenzhen, China
The Hebrew University of Jerusalem, Israel

World Scientific

NEW JERSEY · LONDON · SINGAPORE · BEIJING · SHANGHAI · HONG KONG · TAIPEI · CHENNAI · TOKYO

Published by

World Scientific Publishing Co. Pte. Ltd.

5 Toh Tuck Link, Singapore 596224

USA office: 27 Warren Street, Suite 401-402, Hackensack, NJ 07601

UK office: 57 Shelton Street, Covent Garden, London WC2H 9HE

Library of Congress Control Number: 2023012310

British Library Cataloguing-in-Publication Data
A catalogue record for this book is available from the British Library.

LINEAR ALGEBRA FOR DATA SCIENCE

ISBN 978-981-127-622-4 (hardcover)
ISBN 978-981-127-623-1 (ebook for institutions)
ISBN 978-981-127-624-8 (ebook for individuals)

For any available supplementary material, please visit
https://www.worldscientific.com/worldscibooks/10.1142/13408#t=suppl

Typeset by Stallion Press
Email: enquiries@stallionpress.com

To the memory of my parents-in-law,
Roma and Anshel Schnider

Contents

13. Solutions to Exercises   187

Preface

Linear algebra is linear algebra is linear algebra. So why does the title refer to data science? The answer is that the content of this text is what I believe is what students in data sciences need to know. I tried not to put here what I believe they can do without. Moreover, when exposing students to the notation for vectors and matrices, I am avoiding using physical interpretations such as forces. Best for data sciences students to have in mind an array with figures when they visualize a vector. Likewise, for a matrix.

Having said that, by glancing through the table of content, it is possible to see that the text deals with what appears in many text books: vectors and matrices, linear subspaces, Gram–Schmidt process, projections and least squares, linear functions, systems of linear equations, QR factorization, inverses and pseudo inverses, determinants, eigensystems, symmetric matrices, singular value decomposition and finally stochastic matrices. I would like to mention that complex variables are not dealt with here as I do not find them essential at this stage and may lead to an unnecessary distraction. Good students can fill the gap for more specialized topics in case it is needed at later stages of their studies.

With the exception of Chapter 12, besides high school algebra, there are minimal prerequisites for this text. The text refers a few times to the statistical concepts of (empirical) mean, variance and covariance. There is no need for prior exposure to linear regression. With a few exceptions[a], there is no assumption for any previous knowledge in calculus. Some ease with polynomials is required for Chapter 9. Chapter 12 deals with stochastic

---

[a]Example 3 in Chapter 1, Example 1 in Section 2.4, Section 3.3, a proof in Section 10.1 (which can be skipped) and the use of basic limit results in Chapter 12.

matrices. Some knowledge in elementary probability is required there and admittedly without completing a basic course in probability, the content of this chapter may appear as not having sufficient motivation. It is also the only chapter where some results are left unproved.

This text is based on a course I delivered to second year bachelor students at the Department of Statistics and Data Science at the Hebrew University of Jerusalem in 2019–2020. It is mostly based on what I developed throughout many years of teaching statistics and operations research on what are the essentials in linear algebra for these students, and moreover what are the non-essentials, we can do without. I would like to mention the text [3]. I have learned much from it. Sections 6.1–6.3 and Section 7.1 are heavily based on it.

I was not alone in this project. Much thanks are due to Rachel Buchuk who was my teaching assistant when this course was given. Many of the exercises given here were dealt with in her tutorials. Yonathan Woodbridge was in charge of the solutions given throughout this text. The high quality of his solutions and presentation is clearly noticeable. He also made many comments on the text itself. Zhenduo Wen did a nice work in constructing all the illuminating depictions. Niv Brosh went over the text and made many corrections and suggestions. Special thanks go to Tali Gat-Falik who proof-read the text meticulously. Of course, all errors are mine. Financial support by the Israel Science Foundation (ISF), grant no. 1512/19, is highly appreciated.

Part 1

Vectors

Chapter 1

Vector Algebra

1.1 Definition of Vectors

As in elementary school we need to get used to real numbers, here we need to internalize the concept of vectors. Do not belittle what you have learned in elementary school: the zero ("having nothing") is not a trivial concept, let alone the negative numbers ("having minus two eggs").

We define a *vector* as an array with a number of entries, each of which is a real number. We usually denote vectors by lower case letters such as, u, v or x. The number of entries in the vector is called its *dimension* and in the case where the dimension of v is n, we say that $v \in R^n$. The order in which these n numbers are put matters, and when we write v_i we point to the ith entry, sometimes referred to as the ith *component*.

Example 1. Let

$$v = \begin{pmatrix} 4 \\ -2 \\ 3 \\ 0 \end{pmatrix}.$$

Then, $v \in R^4$ and, for example, $v_1 = 4$.

At times the entries of a vector will be measured with some unit of measurement such as meters, dollars or hours. Although technically possible, sums of the type $v_i + v_j$ when u_i and u_j are measured with different units leads to some meaningless numerical value. On the other hand, in case of a common unit, $u_i + u_j$ is also measured with this unit.

The zero vector is as expected a vector where all its entries are zeros. It is denoted by $\underline{0}$. In general, any vector where all its entries are equal to a constant a, will be denoted by $\underline{a}$. A *unit* (or elementary) vector is a vector where one entry is equal to 1, while the rest of the entries are equal to 0. Among the set of vectors of dimension n, there are n unit vectors. We denote by e_i the ith unit vector, namely the unit vector whose 1 appears at the ith entry, $1 \leq i \leq n$. In short, for an n-entry vector,

$$(e_i)_j = \begin{cases} 1 & 1 \leq j = i \leq n \\ 0 & 1 \leq j \neq i \leq n. \end{cases}$$

For example, the unit vector $e_2 \in R^4$ is

$$e_2 = \begin{pmatrix} 0 \\ 1 \\ 0 \\ 0 \end{pmatrix}.$$

Example 2. A polynomial is a function $P(x) : R^1 \to R^1$, such that for some vector $a = (a_0, a_1, \ldots, a_n) \in R^{n+1}$, $P(x) = \Sigma_{i=0}^n a_i x^i$. Thus, there is a one-on-one correspondence between n-degree polynomials and vectors $a \in R^{n+1}$.

1.2 Vector Operations

We next define some algebra of vectors.

Scalar multiplication. A vector $v \in R^n$ can be multiplied by a real number, referred in this context as a *scalar* α, and hence αv is a vector whose ith entry equals αv_i, $1 \leq i \leq n$. See Fig. 1.1.

Vector summation. Vectors can be summed-up but only if they are of the same dimension: For $u, v \in R^n$, $(u + v)_i = u_i + v_i$, $1 \leq i \leq n$. Otherwise vector summation is not well-defined. Note that $v - u$, which is vector subtraction, is in fact $u + (-1 \times v)$. See Fig. 1.2.

Linear combination. Of course, $(\alpha u + \beta v)_i = \alpha u_i + \beta v_i$, $1 \leq i \leq n$. The vector $\alpha u + \beta v$ is said to be a *linear combination* of u and v with *coefficients* α and β, respectively. Of course, we can add more than two vectors. Also,

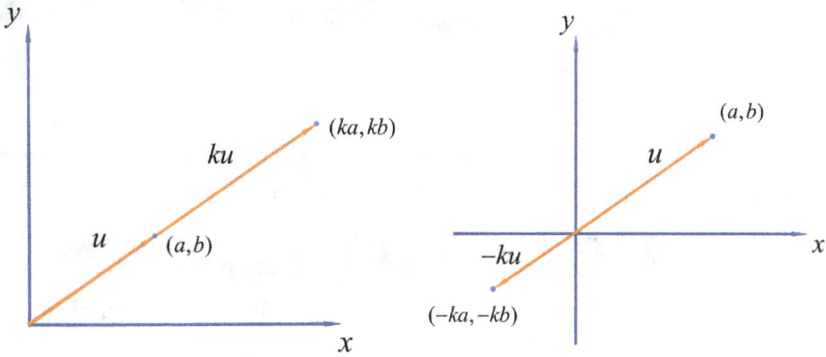

Figure 1.1. Multiplication of a vector by a positive scalar (left), and a negative scalar (right).

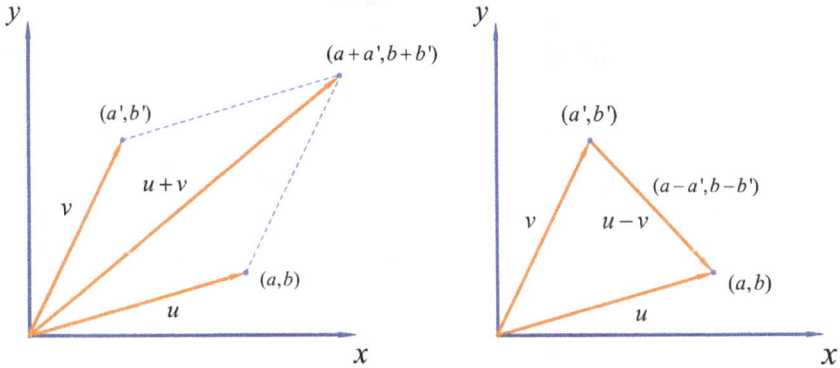

Figure 1.2. Vector summation (left), and subtraction (right).

observe that any vector $v \in R^n$ is a linear combination of the n unit vectors. Indeed,

$$v = \sum_{i=1}^{n} v_i e_i.$$

Exercise 1.1

Define the vectors: $u = (2, -7, 1)$, $v = (-3, 0, 4)$, $w = (0, 5, -8)$.
Calculate: (a) $3u - 4v$ (b) $2u + 3v - 5w$

Exercise 1.2

Find x and y that satisfy:
(a) $(x, 3) = (2, x + y)$ (b) $(4, y) = x(2, 3)$

Inner product. Vectors multiplication is a less obvious concept. We next define the inner (or dot) product between two vectors.

Definition

Definition 1.1. For two vectors $u, v \in R^n$, define a scalar (number) as their inner prod, and denote it by $u^t v$,

$$u^t v = \sum_{i=1}^{n} u_i v_i. \qquad (1.1)$$

Note that the inner product is not well-defined for two vectors who do not have the same dimension. It is easy to see that $(v + u)^t w = v^t w + u^t w$ and that for a scalar a, $(av)^t u = a(v^t u)$.

Example 1 (Descriptive statistics). Let $x \in R^n$. We denote by $\bar{x}$ the (arithmetic) mean of the entries in x, namely $\bar{x} = \frac{1}{n}\Sigma_{i=1}^{n} x_i$. Then, $\bar{x} = \frac{1}{n}x^t \mathbf{1}$. Note that $\underline{\bar{x}} \in R^n$ is the vector with all its entries equal the constant $\bar{x}$. Hence, it is possible to see that

$$\text{Cov}(x, y) = \frac{1}{n}\sum_{i=1}^{n}(x_i - \bar{x})(y_i - \bar{y}) = \frac{1}{n}(x - \underline{\bar{x}})^t(y - \underline{\bar{y}})$$

$$= \frac{1}{n}(x^t y - n\bar{x} \times \bar{y} - n\bar{y} \times \bar{x} + n\bar{x} \times \bar{y})$$

$$= \frac{1}{n}x^t y - \bar{x} \times \bar{y} = \frac{1}{n}x^t y - \frac{1}{n}x^t \mathbf{1} \times \frac{1}{n}y^t \mathbf{1}.$$

In particular,

$$\text{Var}(x) = \text{Cov}(x, x) = \frac{1}{n}(x - \underline{\bar{x}})^t(x - \underline{\bar{x}}) = \frac{1}{n}x^t x - \bar{x}^2$$

$$= \frac{1}{n}x^t x - \frac{1}{n}\left(x^t x - \frac{1}{n}(x^t \mathbf{1})^2\right).$$

Example 2 (Evaluating polynomials). For the polynomial $P(x)$ assume that there exists for some $a \in R^{n+1}$ with $a = (a_0, a_1, \ldots, a_n)$, such that $P(x) = \Sigma_{i=0}^n x^i$. Then, for any scalar x, $P(x) = a^t v$, where $v = (1, x, x^2, \ldots, x^n)$.

Example 3 (Taylor expansion). Let $f(x) : R^n \to R^1$ be a real function. The vector of its partial derivatives, known as the *gradient*, is usually denoted by $\nabla f(x) \in R^n$. Thus, the first order (sometimes called the linear) approximation of $f(x)$ around the point x_0 is

$$f(x) \approx f(x_0) + \nabla f(x_0)^t (x - x_0). \tag{1.2}$$

1.3 Norm of a Vector and Orthogonality

Note that $v^t u = u^t v$ and if $\underline{1}$ is a vector of 1's, then $v^t \underline{1} = \Sigma_{i=1}^n v_i$. Also the mean value of the entries in v is $\frac{1}{n} v^t \underline{1}$. There is a special interest in the square root of the inner product of a vector with itself. It is denoted by $||v||$, called the *norm* of v and is defined as $\sqrt{\Sigma_{i=1}^n v_i^2}$.[a] Note that the norm preserves the unit of measurement of the entries of v: If v_i is measured in meters, $1 \le i \le n$, then the same is the case for $||v||$. Also note that $\frac{1}{\sqrt{n}} ||v - \frac{1}{n} v^t \underline{1}||$ returns SD(v), the standard deviation of the entries of v.

Three properties of the norm are mentioned below. The first two are trivial. The third will be argued for later.

- $||v|| = 0$ if and only if $v = \underline{0}$
- for any scalar α, $||\alpha v|| = |\alpha| ||v||$
- the triangle inequality:

$$||v + u|| \le ||v|| + ||u||. \tag{1.3}$$

Remark. When one looks at the definition of the norm of a vector, one feels that it summarizes in one number its magnitude. This fact is reinforced by the second bullet above: when all entries are, for example, doubled, then the same is the case with the norm. On the other hand, by dividing all the entries of a vector by its norm, one gets a vector whose norm equals one. The resulting vector, called the *normalized vector*, can be said to be without magnitude as only the proportions between the original entries are preserved. Put differently, its entries are free of a unit of measurement. It is hence possible to look at $u/||u||$ as the *direction* of the vector u. In particular, it is possible that two vectors are different (in norm) but equal in

---

[a]Sometimes this norm is called the *Euclidean* norm in order to disassociate it with other norms. There will be no ambiguity in this text so no further reference to Euclidean norm will be made.

their direction. Also, this definition allows size comparison between vectors only when their direction is of concern.

> **Definition**
>
> **Definition 1.2.** Two vectors u and v are said to be orthogonal if $u^t v = 0$.

The following exercises and their solutions appear in [8].

> **Exercise 1.3**
>
> Normalize the following vectors: (a) $u = (-3, 4)$ (b) $v = (4, -2, -3, 8)$ (c) $w = (\frac{1}{2}, \frac{2}{3}, -\frac{1}{4})$

> **Exercise 1.4**
>
> For every $u, v, w \in R^n$ and $k \in R$ prove:
> (a) $(u + v)^t w = u^t w + v^t w$ (b) $(ku)^t v = k(u^t v)$ (c) $u^t v = v^t u$

> **Exercise 1.5**
>
> Find k such that the following vectors are orthogonal: $u = (1, k, -3)$, $v = (2, -5, 4)$.

We next state the famous Pythagoras theorem.

> **Theorem**
>
> **Theorem 1.1.** *The two vectors u and v are orthogonal if and only if*
>
> $$||v + u||^2 = ||v||^2 + ||u||^2.$$

Proof.

$$||v + u||^2 = (v + u)^t(v + u) = v^t v + v^t u + u^t v + v^v$$
$$= v^t v + 2v^t u + u^t u = ||v||^2 + 2v^t u + ||u||^2$$

which indeed equals $||v||^2 + ||u||^2$ if and only if v and u are orthogonal. $\square$

We next state another almost as famous theorem: Cauchy–Schwarz inequality.

> **Theorem**
>
> **Theorem 1.2.** *For any pair of non-zero vectors $u, v \in R^n$,*
>
> $$|v^t u| \leq ||v|| ||u||. \tag{1.4}$$
>
> *Moreover, equality holds if and only if there exists some scalar x such that $v = xu$.*

Proof. First consider the following quadratic function of x:

$$||xu + v||^2 = ||u||^2 x^2 + 2u^t v x + ||v||^2. \tag{1.5}$$

Since it is a non-negative function, we get that $4(u^t v)^2 \leq 4||u||^2 ||v||^2$,[b] which is equivalent to $|v^t u| \leq ||v|| ||u||$, as required. Getting equality means that there exists an x such that $||xu + v||^2 = 0$ or $xu + v = \underline{0}$. This completes our proof. □

Remark. From inequality (1.4) we learn that

$$-1 \leq \frac{v^t u}{||v|| ||u||} \leq 1. \tag{1.6}$$

The right (left, respectively) inequality is tight when u and v come with the same (opposite, respectively) direction, namely $u = \alpha v$ for some $a > 0$ ($\alpha < 0$, respectively). Moreover, the ratio equals zero if and only if the two vectors are orthogonal. Hence, this ratio can be defined as a measure of how close the directions of v and u are. An equivalent measure is to look at

$$\theta = \arccos \frac{v^t u}{||v|| ||u||}, \quad 0 \leq \theta \leq \pi, \tag{1.7}$$

as the *angle* between the two vectors. Indeed, given two vectors, θ is a function only of their directions. Indeed, $\theta = \pi/2$ (90 degrees) if the two vectors are orthogonal, it equals $\theta = 0$ when they share the same direction and $\theta = \pi$ (180 degrees) when their directions are strictly opposing each other.

---

[b]A reminder: a quadratic function $ax^2 + bx + c$ with $a > 0$ is non-negative for any x if and only if $b^2 - 4ac \leq 0$.

Example (Descriptive statistics). Let $x, y \in R^n$. Invoking Inequality (1.6) to the two vectors $u = x - \bar{x}$ and $v = y - \bar{y}$, we conclude that

$$-1 \leq \text{Corr}(x, y) \leq 1,$$

where $\text{Corr}(x, y) = \frac{\text{Cov}(x,y)}{\text{SD}(x)\text{SD}(y)}$. In particular, $\text{Corr}(x, y) = 0$ if and only if $x - \bar{x}$ and $y - \bar{y}$ are orthogonal.

Remark. Assuming both v and u are vectors where the means of their entries equal zero, we can see that Inequality (1.6) implies the well-known fact that the correlation between two variables is a fraction between -1 and 1. As the absolute value of the correlation in invariant with change of scale, this inequality holds for vectors which are not necessarily centered at zero.

We are now ready to prove the triangular inequality.

Proof of Inequality (1.3). We will show the equivalent inequality, that is that

$$||v + u||^2 \leq (||v|| + ||u||)^2.$$

Indeed, the left-hand side equals

$$||v||^2 + 2v^t u + ||u||^2,$$

while the right-hand side equals

$$||v||^2 + 2||v||||u|| + ||u||^2.$$

Invoking Cauchy–Schwarz inequality (see (1.4)), concludes the proof. Note that we can also learn from the above proofs that an equality holds in (1.3) if and only if $v = \alpha u$ for some **positive** α.

The definition of the norm of the difference between two vectors resembles that of a distance between them. Indeed, $||v - u||$ is positive unless $u = v$, in which case it equals zero. This can be seen even further with the following version of the triangle inequality.

Theorem

Theorem 1.3. *For any three vectors $v, u, w \in R^n$,*

$$||u - v|| \leq ||u - w|| + ||w - v||.$$

Proof. Note that

$$v - u = (v - w) + (w - u).$$

The rest now follows from (1.3). □

1.4 Projecting One Vector on Another

Our final concept to be introduced in this chapter is that of a projection of one vector on another.

Definition

Definition 1.3. Assume v and u to be two non-zero vectors in R^n. Let

$$x^* = \arg\min_{x \in R} ||v - xu||. \qquad (1.8)$$

Then $x^* u$ is called the projection of v on (or along) u and it is denoted by $P_u(v)$. The vector $v - x^* u$ is called the residual of the projection.

A hidden assumption in the above definition is that x^* is uniquely defined. This is indeed the case as the following theorem says.

Theorem

Theorem 1.4. x^* *as defined in (1.8) is unique. Moreover, it equals* $\frac{v^t u}{||u||^2}$. *Also, the projection of v on αu is invariant with respect to* $\alpha \neq 0$. *In other words, $P_{\alpha u}(v)$ is not a function of $\alpha \neq 0$. Finally, $P_u(v)$ and $v - P_u(v)$, namely the projection and its residual, are orthogonal.*

Proof. The quadratic function in x, $||v - xu||^2$, equals $||u||^2 x^2 - 2v^t u x + ||v||^2$. This function has a unique minimum at $x^* = \frac{v^t u}{||u||^2}$. Regarding the invariance with respect to $\alpha \neq 0$, by inspecting x^* it is easy to see that as long as $\alpha \neq 0$, the projection of v on αu,

$$\frac{v^t \alpha u}{||\alpha u||^2} \alpha u,$$

is not a function of α. Finally,

$$(v - x^* u)^t x^* u = x^* (v^t u - x^* ||u||^2) = x^* \left(v^t u - \frac{v^t u}{||u||^2} ||u||^2 \right) = x^* 0 = 0,$$

as required. □

From this theorem we can see that if $||u|| = 1$ (which can be assumed without loss of generality as far as the projection and its residual are of concern), we get that the projection of v on u is $(v^t u)u$ and its residual equals $u - (v^t u)u$.

Consider the following two extreme cases. Firstly, if $v = \alpha u$ for some value of $\alpha \neq 0$, we get that v and v's projection on u coincide. The residual then equals the zero vector. In particular, the projection of the projection is the (original) projection, or $P_u(P_u(v)) = P_u(v)$. Secondly and at the other extreme, it is easy to see that the projection equals zero (or more loosely said, does not exist) if and only if v and u are orthogonal. Finally, recall that the norm of the difference between two vectors has the interpretation of the distance between them. Then, $P_u(v)$ is the closest vector to v among all the vectors of the type αu, $\alpha \neq 0$.

Exercise 1.6

Given the vectors $a = (2, 5.1, 7)$, $\quad b = (-3, 6.2, 4)$.

(a) Calculate $||a - \beta b||^2$ for $\beta = 2$ and $\beta = 0.5$.
(b) Find β which minimizes the above norm.
(c) Find the projection of a on b and its residual.

Chapter 2

Linear Independence and Linear Subspaces

2.1 Linear Independence

The concept of linear combination of a number of vectors was defined in the previous chapter. In this chapter we ask whether or not a given vector can be presented as a linear combination of a given set of vectors. Moreover, in the case where the answer is positive, the next question is whether or not the set of corresponding coefficients (the alphas) is unique or not. Towards this end, we need to define the concept of linear independence among a set of vectors.

Theorem

Theorem 2.1. *Let $v^1, v^2, \ldots, v^k$ be a set of k non-zero vectors in R^n. The following are equivalent:*

1. *The unique way to express the zero vector as their linear combination is the trivial one. In other words, if $\Sigma_{i=1}^{k} \alpha_i v^i = \underline{0}$, then $\alpha_i = 0$, $1 \leq i \leq k$.*
2. *None of these k vectors can be expressed as a linear combination of the other $k - 1$ vectors. In other words, for any i, there do not exist $k - 1$ scalars $\alpha_1, \alpha_2, \ldots, \alpha_{i-1}, \alpha_{i+1}, \ldots, \alpha_k$, such that $v^i = \Sigma_{j=1}^{i-1} \alpha_j v^j + \Sigma_{j=i+1}^{k} \alpha_j v^j$, $1 \leq i \leq k$.*

Theorem

Theorem 2.1. *(Continued)*

3. *If a vector $v \in R^n$ can be expressed as a linear combination of this set of vectors, this can be done in a unique way. In other words, if*

$$\sum_{i=1}^{k} \beta_i v^i = \sum_{i=1}^{k} \alpha_i v^i,$$

then $\beta_i = \alpha_i$, $1 \leq i \leq k$.

Proof. Suppose condition 1 does not hold. Then, for some α_i, $1 \leq i \leq n$, not all of which equal to zero, $\underline{0} = \Sigma_{i=1}^{k} \alpha_i v^i$. This implies that there exists at least one (in fact, two) such non-zero α's, say α_i. Then, $-\alpha_i v^i = \Sigma_{j=1}^{i-1} \alpha_j v^j + \Sigma_{j=i+1}^{k} \alpha_j v^j$, or

$$v^i = \sum_{j=1}^{i-1} -\frac{\alpha_j}{\alpha_i} v^j + \sum_{j=i+1}^{k} -\frac{\alpha_j}{\alpha_i} v^j,$$

namely condition 2 does not hold. In this case, we have just stated also a counter-example for condition 3, which hence does not hold: the vector v^i is expressed as two linear combinations of vectors in this set (one of which is trivial: itself). Finally, if condition 3 does not hold then

$$\underline{0} = \sum_{i=1}^{k} (\alpha_i - \beta_i) v^i,$$

thereby violating condition 1, as $\alpha_i - \beta_i \neq 0$ for at least one index for i, $1 \leq i \leq k$.   □

Definition

Definition 2.1. A set of vectors $v^1, v^2, \ldots, v^k$ is said to be linearly independent if one (and hence all) of the conditions detailed in Theorem 2.1 holds.

Note that the vector $\underline{0}$ cannot be a member of a set of linearly independent vectors. Also, if a set of vectors are linearly independent then the same is the case with any subset thereof.

Example 1. Any two non-zero vectors in R^n are linearly independent, unless each one of them is a scalar multiplication of the other.

Example 2. The set of n unit vectors $e_i \in R^n$, $1 \le i \le n$, are linearly independent.

See Figs. 2.1 and 2.2 for an illustration of linear dependence and independence.

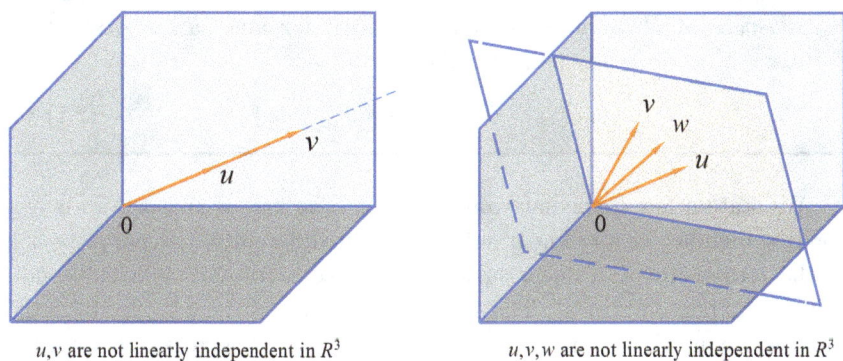

u,v are not linearly independent in R^3 u,v,w are not linearly independent in R^3

Figure 2.1. Not linearly independent vectors.

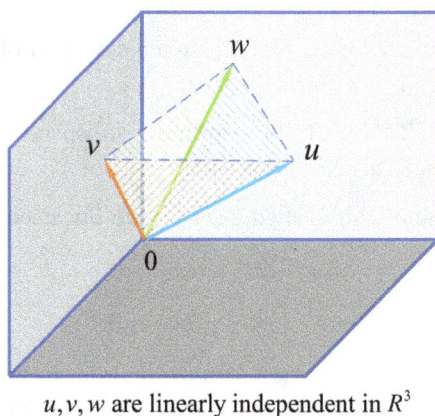

u,v,w are linearly independent in R^3

Figure 2.2. Linearly independent vectors.

2.2   Linear Subspaces

A word on terminology. We introduced below the concept of subspace while not saying first what a space is. The reason is that we try to keep the terminology in line with what is usually used elsewhere but without going through the concept of a space, which for this text is not needed. The same can be said on the word "linear": all subspaces we deal with in the sequel are linear. In doing that we go half-way [3], who refrained from using the terminology of subspaces.

Definition

Definition 2.2. A set of vectors $V \subseteq R^n$ is said to be a *linear subspace* of R^n if it is not empty and if for any pair of vectors $v, u \in V$ and for any pair of scalars α and β,

$$\alpha v + \beta u \in V. \tag{2.1}$$

We state a few immediate observations. The first is that the zero vector is a member of any linear subspace. Put differently, the participation of the zero vector in a linear subspace is a necessary and sufficient condition for its non-emptiness. The second is that unless the zero vector is the only vector in V, V contains an infinite number of vectors. The third is that (2.1) generalizes to any larger linear combination: If $v^1, \ldots, v^k \in V$, then $\Sigma_{i=1}^{k} \alpha_i v^i \in V$ for any set of k scalars α_i, $1 \leq i \leq k$.

Example 1. R^n itself is a linear subspace.

Example 2. Let $v^1, \ldots, v^k \in R^n$. Then the set of all of their linear combinations, $\Sigma_{i=1}^{k} \alpha_i v^i$, is a linear subspace.

Example 3. Let $V^1, V^2 \subseteq R^n$ be two linear subspaces. Then, $V^1 \cap V^2$ is a linear subspace but this is not necessarily so for $V^1 \cup V^2$.

Example 4. Consider the following set of two linear equations:

$$3x_1 - 2x_2 + 4x_3 = 0,$$

$$-x_1 + 3x_2 - 2x_3 = 0.$$

Note that both entries in the right-hand side are zero and this is not coincidental. The first thing to observe is that the zero vector, $(x_1, x_2, x_3) = (0, 0, 0)$ obeys the set of equations, making the solution set not empty. Second, it is a simple exercise to check that if two vectors in R^3 solve them, then the same is the case with respect to any linear combination of them. Thus, the solution set of this set of equations is a linear subspace of R^3. Note that had the right-hand side come with some non-zero entries, this fact would not be true. In particular, the zero vector would not be a solution any more. Thus, having zeros on the right-hand side here is critical for having a linear subspace as its solution set. The importance of this can be deduced from the fact that such a set of linear equations has a special title: *homogeneous equations*. Moreover, any set of equations has its counterpart in homogeneous equations. They are the equations which results from replacing the original right-hand side with a zero right-hand side. For example, consider the set of equations

$$3x_1 - 2x_2 + 4x_3 = 1,$$

$$-x_1 + 3x_2 - 2x_3 = 2.$$

Let $x \in R^3$ be a solution. (we do not claim here that such a solution exists, although this is true for this set: for example, $x = (-1/7, 9/7, 1)$ is a solution.) Then if y is a solution to the corresponding homogeneous equations (for example, $y = (-8/7, 2/7, 1)$), then $x+y$ (for example, $x+y = (-9/7, 11/7, 2)$) also solves them. The converse is also true: if z is another solution to the non-homogeneous set (again we do not claim its existence), then $x - z$ is a solution to the homogeneous version. Needless to say, all stated here holds for any set of linear equations. This is summarized in the next theorem.

> **Theorem**
>
> **Theorem 2.2.** *Let x be a solution to a set of linear equations. Then the same is the case for $x + y$ if and only if y is a solution to its homogeneous version.*

Note that the theorem does not assume the existence of such a solution.

Example 5. Recall the one-on-one correspondence between n-degree polynomials and vectors in R^n, detailed in Example 2 in Section 1.2. It is hence possible to say, when the focus is on the coefficient vectors of the polynomials, that the n-degree polynomials form a linear subspace of R^{n+1}.

Exercise 2.1

Prove that all vectors which are orthogonal to a given vector v form a linear subspace.

Exercise 2.2

Let V_1, V_2 be three linear subspaces of R^n.
(a) Prove that the intersection of V_1 and V_2 form a linear subspace.
(b) Show by a counter-example that the union of V_1 and V_2 does not necessarily form a linear subspace.

2.3 Bases and Dimension

In the case where $V = \{\underline{0}\}$ we define the *dimension* of V as zero. Otherwise, let $v \in R^n$ be a non-zero vector. If you consider all the vectors of the form αv, namely when you run through all possible scalars α, you can see that you have generated a linear subspace.[a] We then say that the resulting linear subspace is *spanned* by v. Suppose now that another non-zero vector $u \in R$ is appended to v. Now consider all vectors of the type $\alpha v + \beta u$. Here too you can observe that a linear subspace is formed. For the case where $u = \alpha v$ for some α this linear subspace coincides with the previous one. Otherwise it is larger and is called the subspace which is spanned by v and u. Note that in the latter case, u and v are linearly independent. Consider now a third non-zero vector $w \in R^n$ and look at all vectors of the shape $\alpha v + \beta u + \gamma w$. Again, if w itself is not already in the linear subspace spanned by v and u, namely there are no α and β such that $w = \alpha v + \beta u$, or put differently, the set of three vectors $\{v, u, w\}$ are linearly independent, a larger linear subspace is formed.

Can we always add more and more vectors and get larger and larger linear subspaces? The answer is no. For example, if you consider the set of unit vectors $\{e_i\}_{i=1}^n$, then any vector in R^n can be expressed as a linear combination of this set. Indeed for any $v \in R^n$, $v = \sum_{i=1}^n v_i e_i$, and hence no extra vectors can be generated via linear combinations when any v is added to the set of unit vectors.

---

[a]Technically speaking, this observation also holds for the zero vector but the generated, trivial, linear subspace is not of our interest and for convenience is excluded.

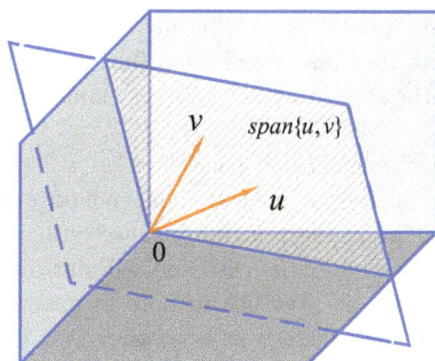

Figure 2.3. Linear span.

Definition

Definition 2.3. Let $v^1, \ldots, v^k$ be a set of non-zero vectors in R^n. Consider the linear subspace

$$V = \left\{ v \in R^n \middle| v = \sum_{i=1}^{k} \alpha_i v^i, \{\alpha_i\}_{i=1}^{k} \right\}.$$

V is called the linear subspace spanned by the $v^1, \ldots, v^k$ and is denoted by $span\{v^1, \ldots, v^k\}$

Note that when considering the above set of vectors, we did not assume that they are linearly independent. Indeed, we can see that if they are not, then at least one vector from this set of vectors can be expressed as a linear combination of the others. This vector can be deleted from this set without harming the capacity of the set to generate vectors via linear combinations. By the same token, we can see that if they are linearly independent, then removing any of the vectors, say v^1, will reduce the set of vectors that can be generated. For example, v^1 cannot be generated by the others and if it is removed from the set $\{v^i\}_{i=1}^{k}$, the set of vectors generated by their linear combination is now reduced. This is not only due to the absenteeism of v^1 itself or any of its scalar multiplications, but also of those generated by adding vectors of the shape αv^1 to all vectors in $span\{v^2, \ldots, v^k\}$.

Of course, even if a linear subspace is defined as the span of a number of vectors, a different set of vectors can span the same linear subspace. For example, v^1 can be replaced by αv^1 for any $\alpha \neq 0$. Also, if v^1 and v^2 are linearly independent, then any one of them (but not both) can be replaced with any vector of the shape $\alpha v^1 + \beta v^2$ for any non-zero pair of coefficients α and β.

Consider again the linear subspace $span\{v^1, \ldots, v^k\}$. As said, if the set of vectors $\{v^i\}_{i=1}^k$ are not linearly independent, one vector from this set can be removed while the remaining subset still spans the same linear subspace. Assume this is the case and without loss of generality the vector to be removed is v^k. We like to note in passing, that if one vector can be removed, then there is at least one other vector which can be removed instead. We can remove any one of them we wish but not necessarily both. Thus, $span\{v^1, \ldots, v^{k-1}\} = span\{v^1, \ldots, v^k\}$. Suppose this process of eliminating vectors goes on and on. Of course, it needs to be ended after a finite number of steps, always leaving the final set not empty. Thus, maybe after reordering the indices of the vectors, for some l, $1 \leq l \leq k$,

$$span\{v^1, \ldots, v^l\} = span\{v^1, \ldots, v^k\},$$

where $\{v^i\}_{i=1}^l$ are linearly independent. This set of vectors is called a *basis* for the linear subspace $span\{v^1, \ldots, v^k\}$. As said, each time the process is iterated, we have some freedom in selecting the vector to be removed. The choices we make affect the terminal set of linearly independent vectors. As we claim below, the size of the terminating set, denoted here by l, does not depend on the selection of the removed vectors and hence it is a property of the linear subspace $span\{v^1, \ldots, v^k\}$ itself. The value of l is called the *dimension* of the linear subspace spanned by a number of vectors.

From now on throughout this text we deal only with linear subspaces which are spanned by a number of vectors and thus for a linear subspace V of dimension ℓ, we denote:

$$\dim(V) = \ell.$$

The terminating set of vectors is said to be the *basis* of $span\{v^1, \ldots, v^k\}$. We, of course, by no means claim that the basis is unique.

Definition

Definition 2.4. A set of vectors $v^1, \ldots, v^k$ is called a basis for a subspace V if they are linearly independent and if $V = span\{v^1, \ldots, v^k\}$ can be expressed as their linear combination.

In other words, a set of linearly independent vectors forms a basis for the subspace they span.

> **Theorem**
>
> **Theorem 2.3.** *Let $\{v^i\}_{i=1}^k$ and let $\{u^i\}_{i=1}^l$ be two sets of linearly independent vectors. Suppose both span (and hence they are bases of) the same linear subspace of R^n. Then $k = l$.*

Proof. Aiming for a contradiction, assume without loss of generality that $k < l$. Consider the vector u^1. Since it is in the linear subspace, it can be expressed as a linear combination of $\{v^i\}_{i=1}^k$. Thus, we can replace one of the v vectors in the former basis by u^1 and keep the resulting set of vectors as a basis. Note that the replaced one can be any vector in the former basis whose coefficient in expressing u^1 as a linear combination of $\{v^i\}_{i=1}^k$ is non-zero. Assume without loss of generality that the replaced vector is v^1. Thus, $(u^1, v^2, v^3, \ldots, v^k)$ is also a basis. Consider now u^2. It can also be expressed as a linear combination of this set. True, it is possible that some of the coefficients in this expression are zeros, but it is impossible that this is the case for all $v^2, v^3, \ldots, v^k$ due to the independence of $\{u^i\}_{i=1}^l$. Thus, u^2 can replace one of the vectors in $v^2, v^3, \ldots, v^k$, say v^2, but still a basis is formed. Thus, $(u^1, u^2, v^3, \ldots, v^k)$ is a basis too. This process can go on and end with $(u^1, u^2, u^3, \ldots, u^k)$ which is hence a basis. In particular, u^{k+1} can be expressed as a linear combination of $(u^1, u^2, u^3, \ldots, u^k)$. This is a contradiction to the assumption that $\{u^i\}_{i=1}^l$ is a basis. □

The following theorem is now immediate.

> **Theorem**
>
> **Theorem 2.4.** $R^n = \text{span}\{e_i, 1 \leq i \leq n\}$. *In particular, the set of elementary vectors forms a basis for R^n and hence its dimension equals n. Moreover, any set of $n + 1$ non-zero vectors in R^n are not linearly independent. Also, any linear subspace of R^n is with a dimension less than or equal to n.*

The set of unit vectors is called the *elementary basis* for R^n. Of course, there are other possible bases. One of interest is $\{f_i\}_{i=1}^n$, where $f_i = \Sigma_{j=1}^i e_j, 1 \leq i \leq n$. For a proof see Exercise B.4.

Note that the definition above does not claim the existence of a basis for any given arbitrary subspace. Yet, this is so far argued for only for subspaces which were defined as the span of a given set of vectors. Toward this end, note that the process of forming a basis stated above can be reversed. Specifically, let $V \in R^n$ be linear subspace. For example, it can be the solution to a set of homogeneous linear equations. Look for a non-zero vector in this subspace and initiate the basis with it. Then look for another vector which is not a scalar multiplication of the first vector and add it to the basis. If such a vector does not exist, you are done. Otherwise, move on and on, each time adding a vector which is not a linear combination of the previous ones. As we argue below this process stops after a finite number of steps. The number of steps is invariant to the vectors which were selected to be added to the set, and as you may recall, it is the dimension of the subspace.

Exercise 2.3

We are given n vectors $x_1, \ldots, x_n \in R^n$ which are all non-zero, such that $x_1, \ldots, x_{n-1}$ are linearly independent. Prove that the vectors $x_1, \ldots, x_n$ are linearly independent if and only if x_n cannot be written as a linear combination of $x_1, \ldots, x_{n-1}$.

Exercise 2.4

Let V, W be linear subspaces of R^n.
(a) Define $V + W = \{v + w | v \in V, w \in W\}$. Prove that $V + W$ is a linear subspace.
(b) Prove that if V and W don't have any common element except for the zero vector, then $\dim(V + W) = \dim(V) + \dim(W)$.

Exercise 2.5

Prove that the set of vectors $\{f_i\}_{i=1}^n$, such that $f_i \in R^n$ and $f_i = \Sigma_{j=1}^i e_j$, $1 \leq i \leq n$, forms a linear basis of R^n.

2.4 Projections on Subspaces

Our final point here deals with projections. You may recall our definition of the projection of a vector on another; see (1.8). As was shown there the projection is invariant with respect to a scalar multiplication of the vector one projects on. Using this section terminology, we can see that when we defined the projection of v on u, we in fact have defined the projection of v on the linear subspace spanned by u. This definition can now be extended to the projection of a vector on any linear subspace.

Definition

Definition 2.5. Let $v \in R^n$ be a vector and let $V \subseteq R^n$ be a linear subspace. Then, the projection of v on V is defined by

$$\arg \min_{u \in V} ||v - u||.$$

In words, it is the closest vector to v among the vectors in V.[b] In particular, if $V = span\{v^i, 1 \leq i \leq k\}$, then the projection of v on V is $\Sigma_{i=1}^{k} \alpha_i^* v^i$, where

$$(\alpha_1^*, \ldots, \alpha_k^*) \in \arg \min_{\alpha_1, \ldots, \alpha_k} ||v - \sum_{i=1}^{k} \alpha_i v^i||.$$

---

[b]We later argue that there exists a unique such vector.

Figure 2.4 illustrates a vector projection. Note that if $\{v^i\}_{i=1}^{k}$ are linearly independent, namely they constitute a basis for V, then $\alpha_1^*, \ldots, \alpha_k^*$ are unique, as we will show below. As expected, $v - \Sigma_{i=1}^{k} \alpha_i^* v^i$ is called the *residual* and the norm of this residual is defined as the *distance* between

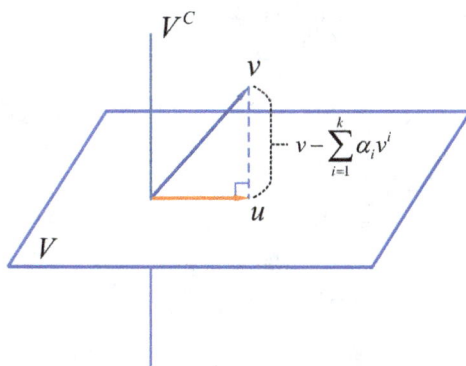

Figure 2.4. Vector projection on the two-dimensional plane.

the vector v and the linear subspace V. Of course, the distance equals zero in the case where $v \in V$. Also, as hinted in Fig. 2.4, the projection and its residual are orthogonal. In fact, the residual is orthogonal to all the vectors in V and not only to the projection. Finding this projection is not a trivial task. We will deal with it, as well as proving all stated here, in the next chapter.

Example 1 (Linear equations). Consider the following set of linear equations, which is called *over determined* as it comes with more equations (constraints) than unknown variables:

$$2x_1 + x_2 = 10,$$

$$x_1 + 2x_2 = 11,$$

$$x_1 + x_2 = 8.$$

Simple inspection shows that a solution does not exist here. For example, you can solve by high school algebra the first two equations and get that uniquely $x_1 = 3$ and $x_2 = 4$. Since $3 + 4 \neq 8$, there is no solution for the whole set of three equations. A second best approach is the *least squares* criterion: look for a pair of x_1 and x_2 that minimizes

$$(10 - (2x_1 + x_2))^2 + (11 - (x_1 + 2x_2))^2 + (8 - (x_1 + x_2))^2. \qquad (2.2)$$

This question is in fact looking for the projection of the vector $(10, 11, 8)$ on the linear subspace spanned by the two vectors $(2, 1, 1)$ and $(1, 2, 1)$. Note that the first vector comes with the constants multiplying x_1, while the second does the same with x_2. Finding the projection goes as follows. Taking derivative of (2.2) once with respect to x_1 and once with respect to x_2 and equating them to zero, leads to the linear equations

$$6x_1 + 5x_2 = 39,$$

$$5x_1 + 6x_2 = 40,$$

which are solved by $x_1 = 34/11$ and $x_2 = 45/11$. Hence, the projection of the right-hand side vector on the subspace spanned by the two columns defining the three linear equations is

$$\frac{34}{11} \begin{pmatrix} 2 \\ 1 \\ 1 \end{pmatrix} + \frac{45}{11} \begin{pmatrix} 1 \\ 2 \\ 1 \end{pmatrix} = \frac{1}{11} \begin{pmatrix} 103 \\ 124 \\ 79 \end{pmatrix}.$$

Note that this projection does not coincide with the right-hand side vector and hence we conclude that a solution to this set of linear equations does not exist.

The casting out algorithm. The above example can be useful when one deals with the issue of finding a basis for the linear subspace defined by a set of vectors which span it. Specifically, let v^i, $1 \leq i \leq m$, be a set of m vectors in R^n and denote by V the linear subspace they span. Our goal is to find a minimal subset among them which spans V as well. The procedure is as follows. Initialize with v^1 being in the basis. Then check if v^1 and v^2 are such that each one of them is a constant multiplication of the other. If the answer is yes, remove v^2. Otherwise, let v^2 be in the basis. From now on (and in fact we could have done that already), assume a number of linearly independent vectors are already in the subset. Then, check whether or not the system of linear equations where one tries to express the next one from this set as a linear combination of those who are already in the basis, is feasible. As seen above this is equivalent to checking whether or not the projection of the next vector on the subspace spanned by the former ones coincides with the candidate-to-enter vector. If the answer is "yes", remove the vector under consideration. Otherwise, enter it to the basis. Note that the above example hints to what one can actually do here: check whether or not a system of linear equations is feasible. How this issue is treated in general is dealt with is an issue we differ to Section 7.1. In any case, move on and consider the next vector. Due to this additional vector, the number of unknowns goes up by one. Note that in the case where the subset comes with n vectors, you can stop: the dimension of $V \subseteq R^n$ cannot exceed n. Also, note that although the end result may depend on the order in which the vectors are considered, this is not the case with regard to the number of vectors there as it coincides with the dimension of V. Finally, note that the casting out algorithm also finds the dimension of V.

2.4.1 *The complementary orthogonal subspace*

Let $W \subset R^n$ be a linear subspace. Denote by W^+ the set of vectors that are orthogonal to all vectors is W. It is easy to check that W^+ is a linear subspace itself, called the *orthogonal linear subspace* of W. Note that $\underline{0} = W \cap W^+$. Above we showed that any vector $v \in R^n$ can be written as $v = w + w^+$ where $w \in W$ and $w^+ \in W^+$. Indeed, w was the projection of v on W and w^+ was its residual. Indirectly, we were able to show that any vector can be written as the sum of two orthoganal vectors, one at a given linear subspace, while the other is in the orthogonal linear space. The next question is if this can be done in a unique way. The answer is positive:

Theorem

Theorem 2.5. *Let W and W^+ be orthogonal linear subspaces with $W \cup W^+ = V$. Then for any $v \in V$ there exists a unique decomposition of v such that $v = w + w^+$, where $w \in W$ and $w^+ \in W^+$.*

Proof. The existence part of the theorem was in fact already established above. Towards uniqueness, let $v = w_1 + w_1^+ = w_2 + w_2^+$ where $w_1, w_2 \in W$ and $w_1,^+ w_2^+ \in W^+$. Hence, $w_1 - w_2 = w_1^+ - w_2^+$. Clearly, $w_1 - w_2 \in W$ and $w_1^+ - w_2^2 \in W^+$. Since they are equal, they belong to $W \cap W^+$. But as we have observed $W \cap W^+ = \underline{0}$. This completes the proof. $\qquad\square$

2.5 Simple Linear Regression

Suppose m individuals were sampled for their weights y_i (in kg) and for their heights x_i (in cm), $1 \leq i \leq m$. In linear regression one looks for an affine function $y = a + bx$ which represents these data well. It is too good to be true that all m points (x_i, y_i), $1 \leq i \leq m$, lie on a single line, so a possible criterion is that of least squares, namely the search for a pair of (a, b) which minimizes the cost function $\Sigma_{i=1}^{m}(y_i - a - bx_i)^2 = ||y - \underline{a} - bx||^2$. Note that a is measured in kg while b is kg/cm. The resulting line is called the *regression line of y on x*. This problem is equivalent to looking for the projection of the vector of $y's$ on the subspace spanned by two linearly independent vectors: one vector with all its entries equal one (and hence it is a constant for every x_i, $1 \leq i \leq n$), namely $\underline{1}$, and another which is the vector of the $x's$. The solution for this optimization problem is as follows.

Theorem

Theorem 2.6. *The regression line $y = a + bx$ is with a slope*

$$b = \frac{\text{Cov}(x, y)}{\text{Var}(x)} = \text{Corr}(x, y)\frac{\text{SD}(y)}{\text{SD}(x)} \qquad (2.3)$$

and an intercept

$$a = \bar{y} - b\bar{x}. \qquad (2.4)$$

Proof. First, a point on notation, as $\bar{y}$ is the scalar which equals the mean of the entries of y, then $\underline{\bar{y}} \in R^m$ is a vector having all its entries equal to $\bar{y}$. A similar definition applies for $\bar{x}$. Then,

$$||y - \underline{a} - bx||^2 = ||y - \underline{\bar{y}} - b(x - \underline{\bar{x}}) - (\underline{a} - \underline{\bar{y}}) - \underline{b\bar{x}}||^2,$$

$$||y - \underline{\bar{y}} - b(x - \underline{\bar{x}})||^2 + ||(\underline{a} - \underline{\bar{y}}) + \underline{b\bar{x}}||^2$$

$$- 2(y - \underline{\bar{y}} - b(x - \underline{\bar{x}}))^t((\underline{a} - \underline{\bar{y}}) + \underline{b\bar{x}}).$$

Since $(y - \underline{\bar{y}})^t \underline{1} = 0$ and $(x - \underline{\bar{x}})^t \underline{1} = 0$, the third entry above is a zero. The second term is non-negative but can be optimized to zero: for any choice of b the corresponding optimal a is $a = \bar{y} - b\bar{x}$. This proves (2.4). Hence, our problem boils down to minimizing the first term. Now,

$$||y - \underline{\bar{y}} - b(x - \underline{\bar{x}})||^2 = ||y - \underline{\bar{y}}||^2 - 2b(y - \underline{\bar{y}})^t(x - \underline{\bar{x}}) + b^2||x - \underline{\bar{x}}||^2.$$

This is a quadratic function in b which is easily seen to be optimized by

$$b = \frac{(y - \underline{\bar{y}})^t(x - \underline{\bar{x}})}{||x - \underline{\bar{x}}||^2},$$

which is the same as (2.3). □

Inspecting (2.4), we observe that the regression line of y on x crosses the point $(\bar{x}, \bar{y})$, which can be looked at as the center of gravity of the data. Hence, an alternative definition of the regression line is the line which crosses this point, and its slope is as stated in (2.3). Finally, minimal algebra leads to the following equivalent statement of the regression line:

$$\frac{y - \bar{y}}{\text{SD}(y)} = \text{Corr}(x, y)\frac{x - \bar{x}}{\text{SD}(x)}.$$

The presentation of the regression line is behind the terminology of regression. Specifically, if an individual comes with a value of x which is k times $\text{SD}(x)$ away from its mean $\bar{x}$, the regression line assigns it a value which is only $\text{Corr}(x, y)k$ times $\text{SD}(y)$ away from its mean $\bar{y}$. In other words, its value is regressed toward the center of the y values in comparison with the position of the x value (both measured in standard deviation units).

We leave this chapter with a few open questions. The first: Given a set of vectors, what is the dimension of the linear subspace they span? In particular, are they linearly independent and if not, which proper subset or in fact subsets of them, form a basis to the linear subspace they all span?

The second: For a vector in the linear subspace spanned by some basis, what are the coefficients of these vectors when this vector is expressed as a linear combination of them? The third: Given a vector which is not in this linear subspace, what is its projection? These questions and more will be dealt with in coming chapters.

Chapter 3

Orthonormal Bases and the Gram–Schmidt Process

3.1 Orthonormal Bases

At the end of the previous chapter we have addressed a number of questions which are of interest but their solutions are not that easy. However, there is a special case where the answer is relatively easy. This is when the basis is formed of orthonormal vectors.

> **Definition**
>
> **Definition 3.1.** A set of k non-zero vectors $\{v^i\}_{i=1}^k$ is said to be orthogonal if $(v^j)^t v^i = 0$ for $1 \leq i \neq j \leq k$. In other words, they are mutually orthogonal. Moreover, they are said to be orthonormal if in addition $\|v^i\| = 1$, $1 \leq i \leq k$.

Since we are interested in issues of linear independence, the linear subspace spanned by a set of vectors, or the projection of a given vector on a linear subspace spanned by a set of vectors, there is no loss of generality in assuming that a set of orthogonal vectors, are in fact orthonormal. Indeed, if one of them does not come with a norm of one, just normalize it by dividing each of its entries with its norm. As we will see below, assuming orthonormality provides much convenience.

> **Theorem**
>
> **Theorem 3.1.** *Suppose the set of non-zero vectors $\{v^i\}_{i=1}^k$ are orthogonal. Then, they are linearly independent.*

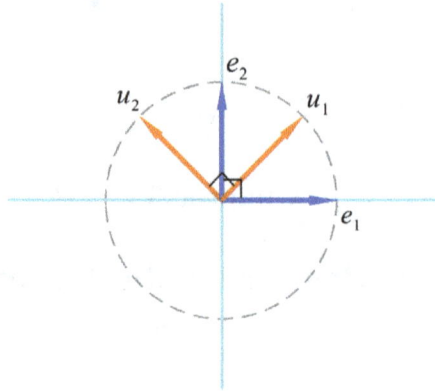

Figure 3.1. Orthonormal basis.

Proof. Aiming for a contradiction, assume that

$$\underline{0} = \sum_{i=1}^{k} \alpha_i v^i,$$

for some scalars α_i, $1 \leq i \leq k$, not all of which are zero. Suppose for the index i $\alpha_i \neq 0$. Then look for the inner product between v^i with the two sides of the above inequality. On the left we of course get the zero scalar. Since v^j and v^i are orthogonal for all $j \neq i$, on the right we get $\alpha_i ||v^i||^2$ which does not equal zero. We have reached a contradiction and the proof is hence completed. $\qquad\square$

The following theorem says that in the case of an orthonormal basis, it is relatively easy to find the coefficients of each vector which is known to belong to the subspace they span. Moreover, it states a condition which determines whether or not a given vector is in the spanned subspace.

Theorem

Theorem 3.2. *Let $\{w^i\}_{i=1}^{k}$ be a set of non-zero orthonormal vectors in R^n and assume that $v \in R^n$ lies in the linear subspace they span, i.e., $v \in span\{w^1, \ldots, w^k\}$. Then,*

$$v = \sum_{i=1}^{k} (v^t w^i) w^i.$$

Proof. By definition,

$$v = \sum_{i=1}^{k} \alpha_i w^i,$$

for some scalars α_i, $1 \le i \le k$. Look for the inner product between w^i and both hand sides of this equality. You should get that

$$(w^i)^t v = \alpha_i \|w^i\|^2, \ 1 \le i \le k.$$

The fact that $\|w^i\| = 1$, $1 \le i \le k$, concludes our proof. $\square$

As the following theorem shows, finding the projection of a given vector on a linear subspace is straightforward when the linear subspace is defined via an orthonormal basis. In fact, the previous theorem is already doing it for the case where the vector and its projection coincide.

Theorem

Theorem 3.3. *Let $\{w^i\}_{i=1}^k$ be a set of non-zero orthonormal vectors in R^n and let $v \in R^n$. Then, the projection of v on the linear subspace spanned by $\{w^i\}_{i=1}^k$ is*

$$\sum_{i=1}^{k} (v^t w^i) w^i. \tag{3.1}$$

Proof. We are in fact asked to solve the following optimization problem:

$$\min_{\alpha_i, i=1,\ldots,k} \left\| v - \sum_{i=1}^{k} \alpha_i w^i \right\|^2.$$

Towards this end, note that

$$\left\| v - \sum_{i=1}^{k} \alpha_i w^i \right\|^2 = \|v\|^2 - 2 \sum_{i=1}^{k} \alpha_i (v^t w^i) + \sum_{i=1}^{k} \alpha_i \sum_{j=1}^{k} \alpha_j (w^i)^t w^j,$$

which by the orthonormality of the vectors $\{w^i\}_{i=1}^k$ reduces to

$$\|v\|^2 - 2 \sum_{i=1}^{k} \alpha_i (v^t w^i) + \sum_{i=1}^{k} \alpha_i^2.$$

Optimizing this function, we can see that it is separable with respect to α_i, $1 \leq i \leq k$, and for each i

$$-2\alpha_i(v^t w^i) + \alpha_i^2, \quad 1 \leq i \leq k$$

can be optimized separately. These functions are quadratic and hence the optimizers are easily seen to equal

$$\alpha_i = v^t w^i, \quad 1 \leq i \leq k,$$

as required. $\square$

Note that Theorem 3.2 in a special case of the result stated in Theorem 3.3 where v lies in $vspan\{w^1, \ldots, w^k\}$.

Recalling the definition of the projection of one vector on another (see Theorem 1.4), we can say that the projection on a linear subspace spanned by orthogonal vectors is the sum of the individual projections of these orthogonal vectors.

By definition $v - \Sigma_{i=1}^k (v^t w^i) w^i$ is the residual of the projection. As expected, the projection $\Sigma_{i=1}^k (v^t w^i) w^i$ and its residual $v - \Sigma_{i=1}^k (v^t w^i) w^i$ are orthogonal. Of course, this property remains regardless of the basis used for $span\{w^1, w^2, \ldots, w^k\}$. In fact, we can state an even stronger result:

Theorem

Theorem 3.4. *Let* $\{w^i\}_{i=1}^k$ *be* k *orthonormal vectors. Then, the residual* $v - \Sigma_{i=1}^k (v^t w^i) w^i$ *and* w^i *are orthogonal for any* i, $1 \leq i \leq k$.

Proof.

$$(w^i)^t \left(v - \sum_{j=1}^k (v^t w^j) w^j \right)$$

$$= (w^i)^t v - (v^t w^i)(w^i)^t w^i = (w^i)^t v - (v^t w^i) = 0, \quad 1 \leq i \leq k.$$
$\square$

3.2   The Gram–Schmidt Process

Due to all of the above we can conclude that all is nice and smooth when the set of vectors are orthogonal. But what can be said in the case where they are not? This is our next issue. What we present is an algorithm whose input is a set of (not necessarily linearly independent) non-zero vectors and whose output is an orthonormal basis for the linear subspace spanned by the input vectors. As a by-product we learn what is the dimension of this linear subspace and in particular if the set of input vectors are linearly independent. Before formally stating the procedure, we first motivate it.

Let v and u be two linearly independent vectors. Recall that $u - \frac{u^t v}{\|v\|^2} v$ is the residual of the of projection of u on v. Consider an arbitrary vector $x \in span(v, u)$. Clearly, for some unique coefficients α and β, $x = \alpha v + \beta u$. Clearly,

$$ x = \left(\alpha + \beta \frac{u^t v}{\|v\|^2} \right) v + \beta \left(u - \frac{u^t v}{\|v\|^2} v \right). $$

In other words, x can be expressed as linear combination of v and the residual of the projection of u on v, two vectors which are orthogonal. In particular, the latter couple of vectors form an orthogonal basis for $span(v, u)$. Of course, they both can be normalized in order to have an orthonormal basis for $span(v, u)$. Note that had v and u not been linearly independent, the residual would have been the zero vector and the second vector in the new basis will turn out to be a zero vector. Indeed, in this case $span(v, u) = span(v)$.

Suppose now that you have three linearly independent vectors v, u and w which, by definition, form a basis for $span(v, u, w)$. Can this basis be replaced with an orthonormal basis? The answer is yes. Specifically, for v and u repeat what was just described above and obtain two orthonormal vectors v_{or} and u_{or} such that $span(v_{or}, u_{or}) = span(v, u)$. Next generate the residual of the projection of w on $span(v_{or}, u_{or})$:

$$ w' = w - (w^t v_{or}) v_{or} - (w^t u_{or}) u_{or}. $$

With the help of Theorem 3.4, it is possible to see that w' is orthogonal to both v_{or} and u_{or} and hence that $span(v_{or}, u_{or}, w) = span(v_{or}, u_{or}, w')$. All that is required now is to normalize w' to $w'/\|w'\|$ in order to get an orthonormal basis. Note that had we have $w \in span(v, u)$, we would get $w' = \underline{0}$. Thus, the procedure outlined here does not only derive an orthonormal basis, but it also generates at least one zero vector when the original set of vectors are not linearly independent. In fact, the dimension of $span(v, u, w)$ equals three minus the number of zero vectors generated. As one can guess, all these can be generalized to more than three vectors given as input. The algorithm is based on the idea that adding a vector to a set of vectors and adding the residual of the projection of this vector on the linear subspace spanned by this set, are equivalent in terms of the new spanned linear subspace. The advantage of the latter option is that it leads, when done recursively, to an orthogonal basis. As a bonus we have in hand a procedure which determines if a set of vectors are linearly independent and the dimension of the linear subspace they span.

The following procedure, known as the Gram–Schmidt process, does exactly that.

The Gram–Schmidt process: General case

Input: $v^1, \ldots, v^m$, m **non-zero vectors**

Output: k, the dimension of $span(v^1, \ldots, v^m)$, and k orthonormal vectors $w^1, \ldots, w^k$ which form a basis to $span(v^1, \ldots, v^m)$.

Initialization: $k = 1$, $w^1 = \frac{1}{\|v^1\|} v^1$

For $i = 2, \ldots, m$ do:

$$x = v^i - \sum_{j=1}^{k} ((v^i)^t w^j) w^j$$

if $x \neq \underline{0}$, then $k \leftarrow k + 1$ and $w^k = \frac{1}{\|x\|} x$

Proof. Note that $\Sigma_{j=1}^{k}((v^i)^t w^j) w^j$ is the projection of v^i on the current $span\{w^1, w^2, \ldots, w^k\}$ and x is its residual. Hence, each time x turns out to be the zero vector in the above procedure, the corresponding v^i vector is a linear combination of the previous vectors $v^1, \ldots, v^{i-1}$ and hence v^i can be dropped from the set of vectors without losing anything in terms of the linear subspace spanned by the original set of vectors. On the other hand, when $x \neq \underline{0}$, this is not the case, and v^i adds capacity in terms of vector generation, leading to an increase in k. The vector x thus replaces v^i in terms of the spanned linear subspace as it is the residual of the projection of v^i on the linear subspace spanned by the previous vectors. In particular, it is orthogonal to all of the previous vectors. Finally, replacing x with w^k is in order to have in hand norm-one vectors. $\square$

Note that the set of k output vector is a function of the way in which the original m vectors are ordered. Of course, the output k is invariant with this order.

For the sake of convenience and later reference we state the procedure for the case where the input vectors are already known to be linearly independent. In this case the dimension of the linear subspace coincides with the number of vectors we have. Moreover, the generate vectors we refer to as x in the process are never the zero vector.

The Gram–Schmidt process: Linear independent case

Input: $v^1, \ldots, v^m$ **non-zero and linearly independent vectors**

Output: m orthonormal vectors $w^1, \ldots, w^m$ that form a basis to $span(v^1, \ldots, v^m)$.

Initialization: $k = 1$, $w^1 = \frac{1}{\|v^1\|} v^1$

For $i = 2, \ldots, n$ do:

$$u^i = v^i - \sum_{j=1}^{i-1} ((v^i)^t w^j) w^j, \tag{3.2}$$

and

$$w^i = \frac{1}{\|u^i\|} u^i. \tag{3.3}$$

Remark. By inspecting (3.2), we can see that if for some j, $1 \le j < i$, v^i and w^j are orthogonal, then nothing is "removed" from v^i in the direction of w^j towards the construction of u^i. The opposite is true when v^i and w^j have close or almost opposite directions, namely if the angle between v^i and w^j (see (1.7) for definition) is close to zero or to π.

Example. Let

$$v^1 = \begin{pmatrix} 1 \\ 1 \\ 1 \end{pmatrix}, \quad v^2 = \begin{pmatrix} 0 \\ 1 \\ 1 \end{pmatrix}, \quad v^3 = \begin{pmatrix} 0 \\ 0 \\ 1 \end{pmatrix}.$$

First,

$$w^1 = -\frac{1}{\sqrt{3}} \begin{pmatrix} 1 \\ 1 \\ 1 \end{pmatrix}.$$

Second,

$$u^2 = \begin{pmatrix} 0 \\ 1 \\ 1 \end{pmatrix} - ((v^2)^t w^1) \frac{1}{\sqrt{3}} \begin{pmatrix} 1 \\ 1 \\ 1 \end{pmatrix} = \begin{pmatrix} 0 \\ 1 \\ 1 \end{pmatrix} - \frac{2}{\sqrt{3}} \frac{1}{\sqrt{3}} \begin{pmatrix} 1 \\ 1 \\ 1 \end{pmatrix} = \frac{1}{3} \begin{pmatrix} -2 \\ 1 \\ 1 \end{pmatrix}$$

and hence

$$w^2 = \frac{1}{\sqrt{6}} \begin{pmatrix} -2 \\ 1 \\ 1 \end{pmatrix}.$$

Finally,

$$u^3 = \begin{pmatrix} 0 \\ 0 \\ 1 \end{pmatrix} - ((v^3)^t w^1) \frac{1}{\sqrt{3}} \begin{pmatrix} 1 \\ 1 \\ 1 \end{pmatrix} - ((v^3)^t w^2) \frac{1}{\sqrt{6}} \begin{pmatrix} -2 \\ 1 \\ 1 \end{pmatrix}$$

$$= \begin{pmatrix} 0 \\ 0 \\ 1 \end{pmatrix} - \frac{1}{\sqrt{3}} \frac{1}{\sqrt{3}} \begin{pmatrix} 1 \\ 1 \\ 1 \end{pmatrix} - \frac{1}{\sqrt{6}} \frac{1}{\sqrt{6}} \begin{pmatrix} -2 \\ 1 \\ 1 \end{pmatrix} = \frac{1}{2} \begin{pmatrix} 0 \\ -1 \\ 1 \end{pmatrix}$$

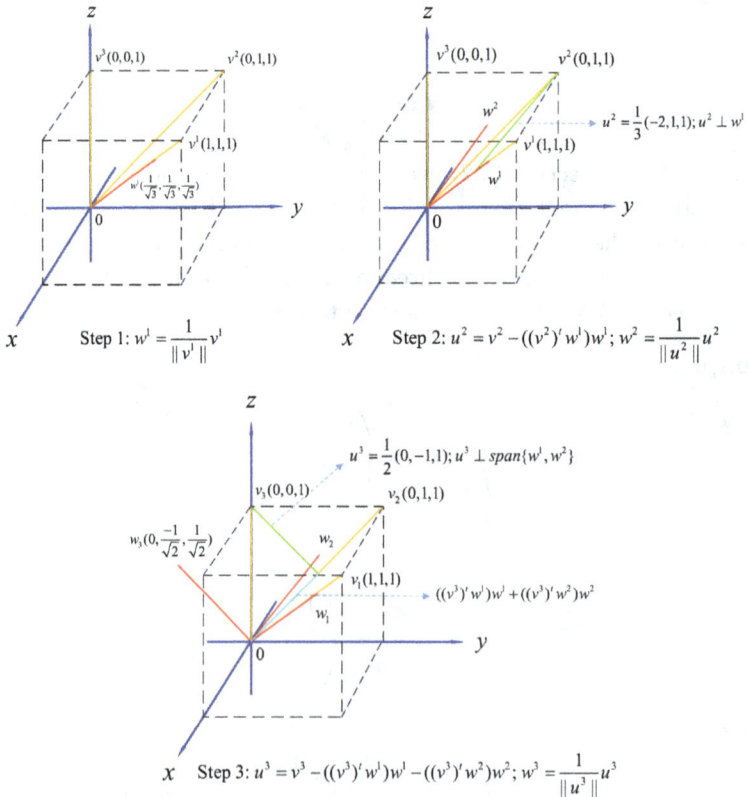

Figure 3.2. The Gram–Schmidt process.

and

$$w^3 = \frac{1}{\sqrt{2}} \begin{pmatrix} 0 \\ -1 \\ 1 \end{pmatrix}.$$

Exercise 3.1

(a) Show that the set of all vectors which are orthogonal to any vector in a linear subspace $V \subset R^n$, is a linear subspace itself. Denote it by V^+.
(b) Show that $(V^+)^+ = V$ (*hint*: use the Gram–Schmidt process).
(c) Show that for any given vector v in R^n, the residual of its projection on V, is the projection on V^+.

Exercise 3.2

Consider R^4 and the subspace $V = span\{v_1, v_2, v_3\}$, where $v_1 = (1, 1, 1, 1)$, $v_2 = (1, 1, 2, 4)$ and $v_3 = (1, 2, -4, -3)$. Find an orthonormal basis for V using the Gram–Schmidt process.

3.3 Optimization Under Equality Constraints

Firstly, a reminder. Let $f(x)$ be a function from R^n to R. Denote by $\nabla f(x) \in R^n$ the vector with the partial derivatives of $f(x)$. A point $x \in R^n$ is called a *stationary point* of $f(x)$ if $\nabla f(x) = \underline{0}$. A necessary condition for a point x_0 to be a local extreme point, be it a minimum or maximum point, is that it is a stationary point.[a] The reason behind that is as follows. Without loss of generality assume that $x_0 = \underline{0}$ and that $f(\underline{0}) = 0$. Then, the linear approximations for $f(x)$ and for $f(-x)$ around $x_0 = \underline{0}$ (see (1.2)) equal

$$f(x) \approx \nabla f(\underline{0})^t x \quad \text{and} \quad f(-x) \approx -\nabla f(\underline{0})^t x,$$

respectively. Thus, unless $\nabla f(\underline{0})^t x = 0$ for all $x \in R^n$, the function $f(x)$ cannot have an extreme point at $\underline{0}$ as otherwise one of the above two values would be positive while the other be negative, making the former an ascend direction (starting at $\underline{0}$) and the latter a descend direction. In particular,

---

[a]The point x_0 is said to be a local minimum if there exists some $\epsilon > 0$ such that for any $x \in R^n$ with $\|x - x_0\| < \epsilon$, $f(x_0) \leq f(x)$. A similar definition exists for local maximum.

$\underline{0}$ would neither be a local maximum point, nor a minimum point. Finally, clearly, $\nabla f(\underline{0})^t x = 0$ for all $x \in R^n$ if and only if $\nabla f(\underline{0}) = \underline{0}$. Indeed, otherwise, namely had $\nabla f(\underline{0}) \neq \underline{0}$, a counter example would be $x = \nabla f(\underline{0})$.

Suppose now that one wishes to find an extreme point for $f(x)$ but under the constraint that $g(x) = 0$ where $g(x)$ is also a function from R^n to R. Note that the zero in the right-hand side is without loss of generality. Suppose x_0 is a feasible point, namely $g(x_0) = 0$, and now the issue is if it is a local extreme point of $f(x)$ or not (among the set of those points meeting the constraint). One can move away from x_0 in many directions but only those that (locally) preserve the constraint are of interest. A first-order approximation for $g(x)$ around x_0 is $g(x_0) + \nabla g(x_0)^t (x - x_0) = \nabla g(x_0)^t (x - x_0)$. Thus, if we ignore second-order terms, we are interested in x, or more precisely, in the direction $x - x_0$, only in the cases where $\nabla g(x_0)^t (x - x_0) = 0$. Looking indeed at $x - x_0$ as a direction denoted by d where a tiny step in this direction does not (practically) violate the constraint, we are interested in directions d which are orthogonal to $\nabla g(x_0)$, namely $\nabla g(x_0)^t d = 0$. Note that the set of such directions form a linear subspace. In particular, if d is such a direction, then the same is the case with $-d$. Other directions are of no interest: In order to get a local constrained extreme point for $f(x)$ at x_0 we need x_0 to be a stationary point for $f(x)$ but only along such directions, directions where locally, the feasibility of x_0 is preserved. Yet, for optimality with respect to the function f, we need $\nabla f(x_0)$ too to be orthogonal to all these directions. This means that $\nabla f(x_0)$ and $\nabla g(x_0)$ need to come with the same (up to a sign) direction, or equivalently put, $\nabla f(x_0)$ needs to be in the linear subspace spanned by $\nabla g(x_0)$.

Suppose now that there are two constraints, $g_1(x) = g_2(x) = 0$. Now the set of directions of interest are those which are in the intersection of the two sets when each of these functions is treated individually as done above. Note that the intersection between two linear subspaces is a linear subspace too but of a reduced dimension. For the latter fact to be strictly true, we assume the technical assumption (called regularity) that $\nabla g_1(x_0)$ and $\nabla g_2(x_0)$ are linearly independent, namely one is not a constant multiplication of the other. Now the condition for a local constrained extreme point is that $\nabla f(x_0)$ is orthogonal to all the directions which are orthogonal to both $g_1(x_0)$ and $\nabla g_2(x_0)$. This means that it belongs to the subspace which is formed by all vectors that are orthogonal to all those which are orthogonal to both $\nabla g(x_1)$ and $\nabla g(x_2)$. Hence, it belongs to the subspace spanned by these two vectors. In summary, a necessary condition for x_0 to be a stationary point is that the gradient of the objective at this point lies in the linear subspace spanned by the gradients of the constrained functions at x_0 (when the latter are assumed to be linearly independent). This is equivalent to requiring the existence of two scalars λ_1 and λ_2 such that

$$\nabla f(x_0) = \lambda_1 \nabla g_1(x_0) + \lambda_2 \nabla g_2(x_0).$$

Moreover, they are unique. In the optimization terminology λ_1 and λ_2 are called *Lagrange multipliers* or *dual variables*.

The above can be extended to the case of m constraints as long as $m \leq n$. The requirement that $m \leq n$ enables the assumption that the m gradients $\nabla g_i(x_0)$, $1 \leq i \leq m$, which are vectors in R^n, are linearly independent. Note that the larger m is, the harder it is to have $g_i(x_0) = 0$, $1 \leq i \leq m$, as a larger set of constraints is concerned. On the other hand, once a point is feasible, then the larger m means the larger is the subspace the gradients of the constraints spanned, making the existence of a set of Lagrangian multipliers more likely. In summary,

---

Theorem

Theorem 3.5. *Assume that x_0 is a feasible point and that $\nabla g_i(x_0)$, $1 \leq i \leq m$, are linearly independent. Then a necessary condition for x_0 to an extreme point, be it a minimum or a maximum, is that $\nabla f(x_0)$ lies in the subspace spanned by $\nabla g_i(x_0)$, $1 \leq i \leq m$. Equivalently, if there exist scalars $\lambda_1, \lambda_2, \ldots, \lambda_m$ such that*

$$\nabla f(x_0) = \sum_{i=1}^{m} \lambda_i \nabla g(x_i). \tag{3.4}$$

Proof. For a formal proof see, e.g., [2, pp. 350–351]. □

Example. Consider the following constrained optimization:

$$\text{optimize}_{x_1,x_2,x_3} f(x) = \frac{1}{2}(x_1^2 + x_2^2 + x_3^2)$$

$$\text{s.t.} \quad g_1(x) = x_1 + x_2 = 1$$

$$g_2(x) = x_2 + x_3 = 1$$

We next prove that $(1/3, 2/3, 1/3)$ is a stationary point. Firstly, it is easy to check that this is a feasible point. Secondly, the gradient of the objective function at this point is $(1/3, 2/3, 1/3)$. Thirdly, the gradient of the first constraint at this point (as at any other point) is $(1, 1, 0)$ and the gradient of the second there is $(0, 1, 1)$. Note that these two gradients are linearly independent. Finally, check that

$$\begin{pmatrix} 1/3 \\ 2/3 \\ 1/3 \end{pmatrix} = \frac{1}{3} \begin{pmatrix} 1 \\ 1 \\ 0 \end{pmatrix} + \frac{1}{3} \begin{pmatrix} 0 \\ 1 \\ 1 \end{pmatrix}.$$

In other words, $\lambda_1 = 1/3$ and $\lambda_2 = 1/3$ are the unique Lagrange multipliers. Note that the key point here is not their uniqueness but their existence.

Chapter 4

Linear Functions

4.1 Definition

Let X and Y be two sets of elements. A function f assigns for any item $\in X$ a (single) item in Y, called $f(x)$. The set X is called the *domain* (of definition) and Y is called the *range*. Note that it is possible to have $y \in Y$ without an $x \in X$ with $y = f(x)$. Also, it is possible for $x_1 \neq x_2$, $f(x_1) = f(x_2)$. The set of all $y \in Y$ such that there exists an $x \in X$ with $y = f(x)$ is called the *image* of the function f and it is denoted by $Img(f)$. We are interested mainly in the case where $X = R^n$ and $Y = R^m$ for some n and m.

Remark. In the case where $m = n$ we can say that when applying the function f to x, then the magnitude of x has been changed from $||x||$ to $||f(x)||$, while its direction has been changed (or rotated) from $x/||x||$ to $f(x)/||f(x)||$.

Definition

Definition 4.1. A function $f : R^n \to R^m$ is said to be linear if for any pair of x_1 and x_2 in R^n and any pair of scalars α and β,

$$f(\alpha x_1 + \beta x_2) = \alpha f(x_1) + \beta f(x_2). \tag{4.1}$$

Examples.

- $f(x) = 3x$ is a linear function from R^1 to R^1.
- Let v be some vector in R^n and for any $x \in R^n$, let $f(x) = v^t x$. Then $f : R^n \to R^1$ is a linear function.

- The following function $f : R^2 \to R^3$ is linear : $f(x_1, x_2) = (3x_1 - 2x_2, 0, -4x_1 + 2x_2)$. Note that if the zero in the right-hand side is replaced by any other real number, the linearity of the function would be violated.
- If $f(x)$ and $g(x)$ are two linear functions from R^n to R^n, then the same is the case with the function $\alpha f(x) + \beta g(x)$ (which equals $f(\alpha x) + g(\beta x)$) for any pair of scalars α and β. Loosely speaking, the set of linear functions is linear itself.
- Projections are linear functions. Inspect (3.1) and notice that

$$\sum_{i=1}^{k} ((\alpha v_1 + \beta v_2)^t w^i) w_i = \alpha \sum_{i=1}^{k} (v_1^t w_i) w^i + \beta \sum_{i=1}^{k} (v_2^t w_i) w^i.$$

Note the following properties. First, the summation in the left-hand side of (4.1) is between two vectors in R^n while the one in the right-hand side is between vectors in R^m. As $f(\underline{0}) = f(\underline{0} + \underline{0}) = f(\underline{0}) + f(\underline{0})$, we conclude that $f(\underline{0}) = \underline{0}$ (but note that the two $\underline{0}$'s here are of different dimensions). In words, any linear function crosses the origin. Indeed, the function from R^1 to R^1, $f(x) = 3x + 2$ is not linear as it does not pass through the origin: $f(0) = 2$ and not 0. In some other contexts it is called a linear function since it represents a line in the two-dimensional plane. In linear algebra circles such a function is called an *affine* function.

Exercise 4.1

Let $V = U = R^3$. Check whether or not $T : U \to V$ is a linear function, where:

$$T \begin{pmatrix} x \\ y \\ z \end{pmatrix} = \begin{pmatrix} x+y \\ 0 \\ 2x+z \end{pmatrix}.$$

Exercise 4.2

Let $V = W = R^2$. Prove or disprove: the function f, which is defined as follows, is linear:

$$f \begin{pmatrix} a \\ b \end{pmatrix} = \begin{pmatrix} a^2 \\ b \end{pmatrix}.$$

4.2 A Linear Function and Its Linear Subspaces

> **Definition**
>
> **Definition 4.2.** The set of vectors $x \in R^n$ for which $f(x) = \underline{0}$ is called the *null set* or the *kernel* of the function f.

> **Theorem**
>
> **Theorem 4.1.** *The null set of a linear function $f : R^n \to R^m$ is a linear subspace of vectors in R^n.*

Proof. Suppose $f(x_1) = f(x_2) = \underline{0}$, namely both x_1 and x_2 are in the null set of f. We need to show that for any pair of scalars α and β, $f(\alpha x_1 + \beta x_2) = \underline{0}$. Indeed, by the linearity of f, $f(\alpha x_1 + \beta x_2) = \alpha f(x_1) + \beta f(x_2) = \alpha \underline{0} + \beta \underline{0} = \underline{0}$. $\square$

Due to this theorem we refer in the case of a linear function to its null set as its null space.

By definition, the domain and the range of a linear function are linear subspaces. The following theorem says that the same is the case with its image.

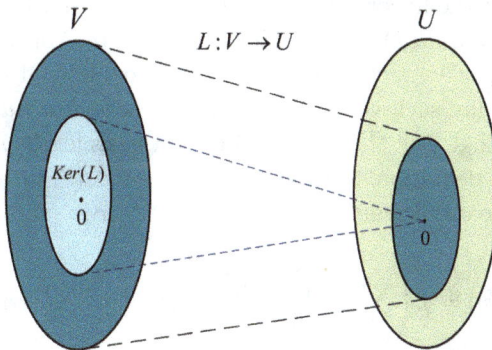

Kernel of a linear function: $L : V \to U$

Figure 4.1. Illustration of the kernel.

Theorem

Theorem 4.2. *Let $f : R^n \to R^m$ be a linear function. Then the set of vectors $y \in R^m$ for which there exists an $x \in R^n$ such that $y = f(x)$, namely $Img(f)$, is a linear subspace of R^m.*

Proof. First, the $Img(f)$ is not empty since $f(\underline{0}) = \underline{0}$. Second, let y_1 and y_2 be two vectors in $Img(f)$. We need to show that the same is the case with $\alpha y_1 + \beta y_2$ for any pair of α and β. Indeed, since y_1 and y_2 are in $Img(f)$, there exist x_1 and x_2 with $y_1 = f(x_1)$ and $y_2 = f(x_2)$. By the linearity of f, we can see that $f(\alpha x_1 + \beta x_2) = \alpha f(x_1) + \beta f(x_2) = \alpha y_1 + \beta y_2$, which implies that $\alpha y_1 + \beta y_2 \in Img(f)$, as required. $\qquad\square$

The following theorem relates the dimensions of the null space and of the image for any linear function:

Theorem

Theorem 4.3. *Let $f : R^n \to R^m$ be a linear function, then*

$$dim(null(f)) + dim(Img(f)) = n.$$

Proof. Let r be $dim(null(f))$ for some, $0 \le r \le n$, and let $v^1, v^2, \ldots, v^r \in R^n$ be a basis for the linear subspace $null(f)$. Note that the basis is empty in the case where $null(f) = \underline{0}$, in which case $r = 0$. Append it with $n - r$ vectors $v^{r+1}, \ldots, v^n \in R^n$ and form a basis for R^n. Of course, there is no need to add any such vectors in the case where $dim(null(f)) = n$. We will next argue that $f(v^{r+1}), \ldots, f(v^n)$ form a basis for $Img(f)$. Indeed, let $u \in Img(f)$. By definition, there exists a vector $v \in R^n$ such that $f(v) = u$. Clearly, for some coefficients $\alpha_1, \ldots, \alpha_n$, $v = \Sigma_{i=1}^n \alpha_i v^i$ and then

$$u = f(v) = f\left(\sum_{i=1}^n \alpha_i v^i\right) = \sum_{i=1}^n \alpha_i f(v^i) = \sum_{i=r+1}^n \alpha_i f(v^i),$$

since $f(v^i) = \underline{0}$ as v^i is in the kernel of $f(\cdot)$, $1 \le i \le r$. In particular, the set of $n - r$ vectors $f(v^{r+1}), \ldots, f(v^n)$ span $Img(f)$. In order to complete the proof, we need show that they are also linearly independent. Aiming for a contradiction, assume that for some $n - r$ coefficients $\alpha_{r+1}, \ldots, \alpha_n$, not all of which are zero,

$$\sum_{i=r+1}^{n} \alpha_i f(v^i) = \underline{0}.$$

Thus, $f(\Sigma_{i=r+1}^{n}\alpha_i v^i) = \underline{0}$, namely $\Sigma_{i=r+1}^{n}\alpha_i v^i$ is in the $null(f)$. This contradicts the way the vectors $v^{r+1}, \ldots, v^n$ were defined. $\qquad\square$

Consider a linear function: $f : R^n \to R^m$. For the n unit vectors $e_i \in R^n$, denote $f(e_i)$ by f_i, $1 \le i \le n$. Let $x \in R^n$ be an arbitrary vector. Of course, some scalars x_i, $1 \le i \le n$, $x = \Sigma_{i=1}^{n} x_i e_i$. Then, by the linearity of f,

$$f(x) = f\left(\sum_{i=1}^{n} x_i e_i\right) = \sum_{i=1}^{n} x_i f(e_i) = \sum_{i=1}^{n} x_i f_i.$$

In other words, $f(x)$ preserves in $Img(f)$ the linear combination expressing x in the domain. As importantly, once a linear function assigns points (which can be of one's choice) in the range for the n unit vectors, there are no more degrees of freedom in constructing the function for the other vectors: they are determined by the former finite group of n vectors. Note that the same observation holds for any basis for R^n and not only for the elementary basis. Stated differently, we can say that if two linear functions agree when applied to a basis in the domain, they are in fact the same function.

Suppose now that $m = n$, namely the domain and the range coincide. From Theorem 4.3 we learn that $dim(Img(f))$ equals n too if and only if $dim(null(f)) = 0$, namely $f(x) = \underline{0}$ if and only if $x = \underline{0}$. We can also see then that $f(x) = f(y)$ if and only if $x = y$ as otherwise $\underline{0} \ne x - y \in null(f)$ too. It is also possible to argue that for any $y \in R^n$ there exists a unique x such that $y = f(x)$: otherwise, $dim(Img(f)) < n$. In this way we can assign for any vector in $Img(f)$ a (now unique) vector in the domain. This function is called the *inverse* of f and is denoted by f^{-1}. We repeat that this inverse function exists if and only if $null(f) = \underline{0}$. Note that $(f^{-1})^{-1} = f$ and if we apply the function f^{-1} after applying f, we in fact apply the identity function as any $x \in R^n$ lands back on x. Finally, note that if f is linear, then same is the case with f^{-1}. We leave the proof of this final fact to the reader.

Our final point here concerns compositions of linear functions. Let $f : R^n \to R^m$ and let $g : R^m \to R^k$ be two linear functions. We denote by $g \circ f$ the function from R^n to R^k which is defined by $g \circ f(x) = g(f(x))$.

In words, take $x \in R^n$, apply the function f on it to get $f(x) \in R^m$. Then, on this vector apply the function g. Note that the order in which the composition is applied matters and in fact it is possible that $g \circ f$ is well-defined while $f \circ g$ is not. The latter case occurs $g(x)$ for some x in the domain of g is not in the domain of f.

> **Theorem**
>
> **Theorem 4.4.** *Let f and g be two linear functions, then the same is the case with $g \circ f$ (which is assumed to be well defined).*

Proof. First, note that $g \circ f(\underline{0}) = \underline{0}$. Second, for any pair of scalars α and β,

$$(g \circ f)(\alpha x + \beta y) = g(f(\alpha x + \beta y)) = g(\alpha f(x) + \beta f(y))$$
$$= \alpha g(f(x)) + \beta g(f(y))$$
$$= \alpha (g \circ f)(x) + \beta (g \circ f)(y).$$ $\qquad\square$

Note that if f^{-1} exists then by definition both $f^{-1} \circ f$ and $f \circ f^{-1}$ exist and are equal to the identity function.

Part 2

Matrices

Chapter 5

Matrices and Matrix Operations

5.1 Basic Concepts

A *matrix* is a rectangular array having a number of rows and columns with real numbers as its entries. We usually refer to matrices by upper case letters. For example, the matrix

$$A = \begin{pmatrix} 2 & -1 & 5 & 7 \\ 3 & 0 & -4 & 8 \\ -1 & 7 & 0 & 9 \end{pmatrix}$$

has three rows and four columns. We say that $A \subset R^{3 \times 4}$. We refer by A_{ij} to the number in the ijth entry, namely the one in row i and column j. For example, $A_{23} = -4$. A vector is in fact a matrix with one column and a scalar is a matrix with one row and one column.

A matrix $A^{m \times n}$ is said to be *tall* (*wide*, respectively) if $m \geq n$ ($m \leq n$, respectively). A matrix which is both tall and wide is called a *square* matrix. Of course, in this case $n = m$. For any matrix $A \in R^{m \times n}$, we refer as its *transpose* matrix, and denote it by A^t. A^t is the matrix in $R^{n \times m}$ and $A^t_{ji} = A_{ij}$, $1 \leq i \leq m$, $1 \leq j \leq n$. A square matrix is said to be *symmetric* if $A^t = A$. A square matrix $A \in R^{n \times n}$ is said to be *upper-triangular* if $A_{ij} = 0$ when $1 \leq j < i \leq n$. It is said to be *lower triangular* if A^t is upper-triangular. Finally, it is said to be *diagonal* if it is both upper and lower-triangular. Note that $A \in R^{n \times n}$ is diagonal if $A_{ij} = 0$ for $1 \leq i \neq j \leq n$.

We next state two matrix operations.

Scalar multiplication. For a scalar a and matrix $A \in R^{m \times n}$, we define the matrix $B \in R^{m \times n}$ by $B = aA$ where $B_{ij} = aA_{ij}$, $1 \leq i \leq m$, $1 \leq j \leq n$.

Matrix summation. For the matrices $A, B \in R^{m \times n}$ we define the matrix $A + B \in R^{m \times n}$ via $(A + B)_{ij} = A_{ij} + B_{ij}$. Note that the dimensions of A coincide with the dimension in B. Otherwise, matrix summation is not well-defined. Note that $B + A = A + B$ (a matrix identity) and $aA + aB = a(A + B)$ (ditto). Later we state the rationale behind these two definitions (although they both seem natural, and at first glance one wonders why justification is called for here).

5.2   Matrix by Vector Multiplication

Let $A \in R^{m \times n}$ and $x \in R^n$. Note that the number of columns in A is equal to the number of entries in the x. We denote by Ax, and call it "A times x", the vector in R^m (recall that m is the number of rows in A), where $(Ax)_i$ is defined via

$$(Ax)_i = \sum_{j=1}^{n} A_{ij} x_j, \quad 1 \leq i \leq m.$$

Note that Ax is the linear combination of the columns of A with coefficients given in x. Indeed, $Ax = \Sigma_{i=1}^{m} x_i A_{.i}$, where $A_{.i}$ is the ith column of A, $1 \leq i \leq m$. Also, when we look at the system of linear equations $Ax = b$ we are first looking if b is in the linear subspace spanned by the columns of A. If this is the case, the solution x, which might or might not be unique, means what are the coefficients of these columns when b is expressed as their linear combination.

The following are immediate:

- For scalars α and β, vectors $x, y \in R^n$ and matrix $A \in R^{m \times n}$, $A(\alpha x + \beta y) = \alpha Ax + \beta Ay$. Also, $A\underline{0} = \underline{0}$. In other words, matrix by vector multiplication is a linear function.
- For two matrices $A, B \in R^{m \times n}$, any two scalars α and β and vector $x \in R^n$, $(aA + \beta B)x = \alpha Ax + \beta Bx$. In words, the linearity of the two matrix operations, Ax and Bx, is preserved under a linear operation on the matrices themselves. Turning this fact on its head, it is this linearity property which we wanted to preserve that led in fact to the way scalar by matrix and matrix by vector multiplications were defined.
- Let $I \in R^{n \times n}$ be a diagonal matrix with $I_{ii} = 1$, $1 \leq i \leq n$. Then for any $x \in R^n$, $Ix = x$. No wonder, I is called the *identity* matrix. Note that for integer n there is an identity matrix of the corresponding size.

- Let $x, y \in R^n$. Looking at x as a matrix, $x \in R^{n \times 1}$. Its transpose is the matrix $x^t \in R^{1 \times n}$. Its matrix by vector multiplication with y, $x^t y$, yields the scalar $\sum_{i=1}^n x_i y_i$ which is the inner product between x and y. Note that the notation used here coincides with the notation used when we defined the inner product between two vectors. We can also observe that $y^t x = x^t y$.

As already said, for $A \in R^{m \times n}$, Ax is a linear function. What we show next is the converse of this fact.

> ### Theorem
>
> **Theorem 5.1.** *Let $f : R^n \to R^m$ be a linear function. Then, there exists a unique matrix $A \in R^{m \times n}$ such that $Ax = f(x)$ for any $x \in R^n$. Moreover, in such a matrix, its jth column is $f(e_j)$, $1 \le j \le n$.*

Proof. It was already said in Chapter 4 that if two linear functions from R^n to R^m agree when applied to n vectors that form a basis for R^n, they agree elsewhere as well, and hence they are in fact the same function. It is easy to see that Ae_j yields the jth column of A, so if we insert $f(e_j)$ in the jth column of A, we will get that $Ae_j = f(e_j)$, $1 \le j \le n$, as required. The uniqueness follows from the fact that if $Ax = Bx$, and hence $(A - B)x = \bar{0}$ for all vectors x (with the required dimension), then $A - B$ is the zero vector, implying that $B = A$. □

Exercise 5.1

Let $A \in R^{m \times n}$ and $f : R^n \to R^m$, such that the vector $f(x)$ is defined as $(f(x))_i = \sum_{j=1}^n A_{ij} x_j$. Prove that $f(x)$ is a linear function.

Exercise 5.2

The following function is linear (see exercise 4.1):

$$T \begin{pmatrix} x \\ y \\ z \end{pmatrix} = \begin{pmatrix} x+y \\ 0 \\ 2x+z \end{pmatrix}.$$

Find the matrix A which represents the above function. For that purpose, calculate $T(e_i)$ for $1 \le i \le 3$, and calculate $\sum_{i=1}^3 x_i T(e_i)$.

Exercise 5.3

(a) Prove that the residual of the projection of x on b is a linear function.
(b) Find the representative matrix A of the linear function specified in (a).

Exercise 5.4

Suppose that $F(x) : R^n \to R^n$ satisfies $F_i(x) = x - \bar{x}$, such that $\bar{x} = \frac{1}{n} \sum_{i=1}^{n} x_i$.
(a) Prove that $F(x)$ is a linear function.
(b) What is the representative matrix A of the function $F(x)$, with respect to the elementary basis?
(c) What is $dim(null(F))$?

5.3 The Rank of a Matrix

Let $A \in R^{m \times n}$. By definition, the dimension of the range of Ax is m. What is the dimension of its image? Note that a vector $y \in R^m$ is in this image if and only if there exists an $x \in R^n$ such that $Ax = y$, namely y is some linear combination of the n columns of A. Hence, the dimension of this image is as the number of linearly independent columns among the n columns of A. Thus, if these columns are the input of the Gram–Schmidt process, one will get as output the dimension of this linear subspace. A similar question can be asked about A^t. Now the input will be the rows of A and the output will tell the number of linearly independent rows (since now the rows of the matrix A are looked at as vectors in R^n) among the m rows of A. The following is an interesting result.

Theorem

Theorem 5.2. *The number of linearly independent columns and the number of linearly independent rows in a matrix are equal.*

Proof. Let $A \in R^{m \times n}$. Denote by $r \leq m$ the number of linearly independent rows in A. We will establish the theorem by showing that there exist r vectors such that each of the columns of A is a linear combination of these r vectors. This implies that the dimension of the linear subspace spanned by the rows is greater than or equal to that spanned by the columns, but since we can swap the roles of rows and columns, an equality in fact exists here.

Let $S_1, \ldots, S_r$ be a set of linearly independent rows. Write each of the m rows of A, denoted here by $R_1, \ldots, R_m$, as their linear combination. Thus, for some coefficients k_{ij}, $1 \leq i \leq m$, $1 \leq j \leq n$,

$$R_1 = k_{11}S_1 + k_{12}S_2 + \cdots + k_{1r}S_r$$

$$R_2 = k_{21}S_1 + k_{22}S_2 + \cdots + k_{2r}S_r$$

$$\vdots$$

$$R_m = k_{m1}S_1 + k_{m2}S_2 + \cdots + k_{mr}S_r$$

Look now at the jth entry across all vectors above, and note that the values on the left hand sides are entries in the jth column of A. Specifically, for j, $1 \leq j \leq n$,

$$A_{1j} = k_{11}S_{1j} + k_{12}S_{2j} + \cdots + k_{1r}S_{rj}$$

$$A_{2j} = k_{21}S_{1j} + k_{22}S_{2j} + \cdots + k_{2r}S_{rj}$$

$$\vdots$$

$$A_{mj} = k_{m1}S_{1j} + k_{m2}S_{2j} + \cdots + k_{mr}S_{rj}$$

Observe now that this column of A is expressed as a linear combination of r vectors, the first vector among them being $(k_{11}, k_{21}, \ldots, k_{m1})^t$, the second being $(k_{12}, k_{22}, \ldots, k_{m2})^t$, etc. As this can be done for any column, the proof is completed. $\square$

The common value addressed in the above theorem is called the *rank* of A and is denoted by $rank(A)$. A matrix $A \in R^{m \times n}$ is said to have a *full rank* if $rank(A) = \min\{m, n\}$. Also, denote by $null(A)$ the dimension of the null-space of the linear function Ax.

Recall that for a matrix $A \in R^{m \times n}$, the operation $f(x) = Ax$ is a linear function $f : R^n \to R^m$. The dimension of $Img(f)$ coincides with the number of linearly independent columns of A, namely with $rank(A)$. Invoking Theorem 4.3, we conclude that:

Theorem

Theorem 5.3.

$$dim(null(A)) = n - rank(A).$$

5.4 Linear Equations and Homogeneous Equations

A linear set of equations with m equations and n unknowns can be defined as $Ax = b$ for some matrix $A \in R^{m \times n}$ and a vector $b \in R^m$, the latter sometimes referred to as the *right-hand side vector*. Since Ax is in fact a linear combination of the columns of A, where x states the corresponding coefficients. The existence of a solution means that b lies in the linear subspace spanned by the columns of A. Moreover, any solution consists of possible coefficients defining a linear combination of these columns which yields b. If the columns of A are linearly independent and a solution exists, then the set of coefficients needs to be unique. Note that the converse is also true: Any set of linear equations can be presented in the form of $AX = b$ for some matrix A and a right-hand side vector b.

For any set of linear equations $Ax = b$, there exists a corresponding set of linear equations which are referred to as a *homogeneous* set of equations: $Ax = \underline{0}$. All said above for any vector b holds for the homogeneous case. But there is something which is unique for $\underline{0}$ as the right-hand side vector: its solution set is never empty. Moreover, it forms a linear subspace of R^n. This observation is easily seen so we omit a proof. The following theorem appeared already as Example 4 in Section 2.2 and is repeated here for convenience.

> **Theorem**
>
> **Theorem 5.4.** *Consider the linear system of equations $Ax = b$. Assume $Ax^* = b$. Then $Ax = b$ if and only if $A(x - x^*) = \underline{0}$.*

Proof. Firstly, observe that if $Ax^* = b$ and $Ax = b$. Then, $A(x - x^*) = Ax - Ax^* = b - b = \underline{0}$. Secondly and for the converse, suppose $Ax^* = b$ and $A(x - x^*) = \underline{0}$. Then, $A(x^* + (x - x^*)) = Ax^* + A(x - x^*) = b + \underline{0} = b$. $\qquad \square$

See Fig. 5.1 for illustration of the general solution of $AX = b$.

5.5 Matrix by Matrix Multiplication

Matrix summation defined above seems natural and maybe the one we would apply without much agony. However, recall that this definition preserves the linearity among linear functions themselves. When it comes to

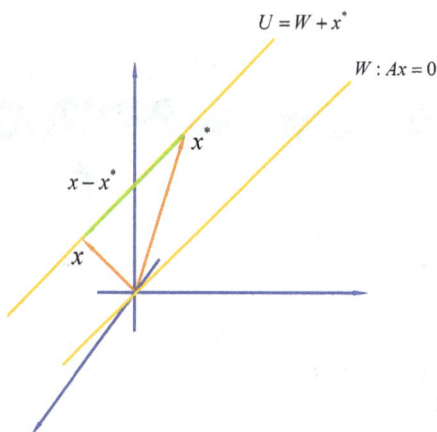

$U = W + x^*$

$W : Ax = 0$

x^*

$x - x^*$

x

x

x^* be a particular solution of $Ax = b$

W be the general solution of $Ax = 0$

$U = x^* + W = \{x^* + w : w \in W\}$ be the general solution of $Ax = b$

Also, the subtraction of two solutions x and $x^* : x - x^* \in W$, that is, $A(x - x^*) = 0$

Figure 5.1. The general solution of $Ax = b$.

matrix multiplication things are less obvious but the definition of matrix multiplication is based on the same idea.

> ## Definition
>
> **Definition 5.1.** Let $A \in R^{m \times k}$ and $B \in R^{k \times n}$. The matrix $C \in R^{m \times n}$ which is called A times B, denoted in short by AB, is defined by
>
> $$C_{ij} = (AB)_{ij} = \sum_{\ell=1}^{k} A_{i\ell} B_{\ell j}, \quad 1 \le i \le m, \ 1 \le j \le n.$$

Note that the ijth entry in the product of A and B is the inner product between the ith row of A and the jth column of B. The first thing to observe is that the product between two matrices is well defined if and only if the number of columns of the left matrix is equal to the number of rows of the right. In particular, it is possible that AB is well defined but BA is not. Moreover, we do not have the commutative property: It is possible that both AB and BA are well defined but $AB \neq BA$. In particular, the dimensions of AB and BA might be different. Finally,

---

Theorem

Theorem 5.5.

$$(AB)^t = B^t A^t. \tag{5.1}$$

Proof. The first thing to observe is that the product AB is well defined if and only if the same is the case with $B^t A^t$. The details are left to the reader. Then, for all pairs of i and j,

$$(AB)^t_{ij} = (AB)_{ji} = \sum_k A_{jk} B_{ki} = \sum_k A^t_{kj} B^t_{ik} = \sum_k B^t_{ik} A^t_{kj} = (B^t A^t)_{ij}.$$
□

Example 1. Note that $A^t A$ and AA^t are well defined. Also note that both of them are symmetric matrices. Indeed, $(A^t A)^t = A^t (A^t)^t$ (by (5.1)), which is back $A^t A$. Yet, unless A is a square matrix, the products lead to symmetric matrices of different dimension. Note that A being square does not guarantee that the two product coincide and usually they do not. They do when A is symmetric.

Example 2 (Inner and outer product between vectors). Let v and u be two non-zero vectors in R^n. Considering them as matrices, we can say that $v, u \in R^{n \times 1}$. Consider the matrix multiplication $v^t u$. It is well-defined and it is a 1×1 matrix. In fact, it is the scalar $\sum_{i=1}^n v_i u_i$. Trivially, the rank of this matrix is one. This coincides with the inner product between two vectors as defined in (1.1). Note that vu^t is also well defined. Moreover, $vu^t \in R^{n \times n}$ where

$$(vu^t)_{ij} = v_i u_j, \quad 1 \leq i, j \leq n.$$

Note that all columns of vu^t are scalar multiplications of v. The same can be said on the rows (when looked at as vectors), now with respect to u. Thus, both linear subspaces spanned by the columns and by the rows of this matrix have a dimension of one. In particular, vu^t is a rank-one matrix.

Example 3 (The identity matrix). You may recall the identity matrix, I. It was defined as the diagonal (square) matrix whose all diagonal entries equal one. It was called this way since for every vector x, if Ix is well defined, then $Ix = x$. The same can be said on matrix multiplication: If AI and/or IA are well defined, then they equal A.

Example 4. The distributive law holds here:

$$A(B + C) = AB + BC.$$

The proof of this fact is left for the reader.

The following theorem is in fact the justification of the way matrix multiplication is defined.

Theorem

Theorem 5.6. *Let* $g : R^k \to R^m$ *and* $f : R^n \to R^k$ *be two linear functions. Note that the range of* f *coincides with the domain of* g. *In particular, the composite linear function* $g \circ f : R^n \to R^m$ *is well defined. Let* $A \in R^{m \times k}$, $B \in R^{k \times n}$ *and* $C \in R^{m \times n}$ *be the matrices which correspond to the linear functions* g, f *and* $g \circ f$, *respectively. Specifically,* $g(x) = Ax$, $f(x) = Bx$ *and* $(g \circ f)(x) = Cx$ *for any* $x \in R^n$. *Then,* $C = AB$. *Put differently, for any* $x \in R^n$,

$$(AB)x = A(Bx).$$

Proof. We need to show that for any $x \in R^n$, $[(AB)x]_i = [A(Bx)]_i$, $1 \le i \le m$. Indeed,

$$[(AB)x]_i = \sum_{j=1}^{n}(AB)_{ij}x_j = \sum_{j=1}^{n}\sum_{l=1}^{k}(A_{il}B_{lj})x_j$$

$$= \sum_{l=1}^{k}A_{il}\sum_{j=1}^{n}B_{lj}x_j = \sum_{l=1}^{k}A_{il}(Bx)_l = [A(Bx)]_i, \quad 1 \le i \le m,$$

as required. $\square$

Example 5 (The matrix of the projection operation). Recall that the $\Sigma_{i=1}^{k}(v^t w_i)w^i$ is the projection of the vector v on the subspace spanned by the orthonomal set of vectors w_i, $1 \le i \le k$. See (3.1). It is possible to see that $(v^t w_i)w^i = w^i w_i^t v$, $1 \le i \le k$. Hence, the projection equals $\Sigma_{i=1}^{k}w^i w_i^t v$, which is turn equals $(\Sigma_{i=1}^{k}w^i w_i^t)v$. In summary, for the matrix $A = \Sigma_{i=1}^{k}w^i w_i^t$, which is the sum of k rank-one matrices, Av yields the projection. Note that A is a square matrix. Moreover, since the projection of the projection is the original projection, $A^2v = A(Av) = Av$. From here the fact that $A^m = A$, for any $m \geq 1$, follows.

Theorem

Theorem 5.7. *Matrix multiplication is associative: for any A, B and C, when well defined,*

$$(AB)C = A(BC).$$

The proof is left as an exercise. Due to this theorem we can write the product between three matrices as ABC, without the need to resort to parenthesis. A corollary here is that, for example, for any square matrix, $A^4 = (A^2)^2$ so A^4 can be computed with two, instead of three, matrix multiplication. In general, A^{2^k} can be computed in k matrix multiplication, instead of $2^k - 1$ when done one by one. Put differently, computing A^k requires only $\log_2 k$ matrix multiplications. Finally, Theorem 5.7 can be generalized for the product among more than two matrices: $(ABC)^t = C^t B^t A^t$, etc.

Exercise 5.5

Let $A = \left(\begin{smallmatrix} 4 & 1 & 0 \\ 5 & 8 & 3 \end{smallmatrix}\right)$ and $B = \left(\begin{smallmatrix} 1 & 9 & 3 & 1 \\ 7 & 2 & 8 & 1 \\ 4 & 0 & 6 & 5 \end{smallmatrix}\right)$. Compute AB.

Exercise 5.6

Suppose that $A \in R^{m \times k}$ and $B \in R^{k \times n}$.
(a) For $C \in R^{k \times n}$, prove that $A(B + C) = AB + AC$.
(b) For $C \in R^{n \times p}$, prove that $(AB)C = A(BC)$.

Exercise 5.7

For $A, B \in R^{m \times n}$ and $C \in R^{n \times k}$ prove that:
(a) $(A + B)C = AC + BC$.
(b) $(A^t B)C = A^t(BC) = A^t BC$.

5.6 The QR Factorization

Let $A \in R^{m \times n}$ be a tall matrix. Also assume that $rank(A) = n$, namely it is a full-rank matrix, which also means that its n columns are linearly independent. For this context, we denote its columns by $v^i \in R^m$, $1 \leq i \leq n$. Applying the Gram–Schmidt process on these columns (see Section 3.2), results in n orthonormal vectors, $w^i \in R^m$, $1 \leq i \leq n$. From (3.2) and (3.3) we learn that

$$Fv^i = \sum_{j=1}^{i-1}((v^i)^t w^j)w^j + ||u^i||w^i, \quad 1 \leq i \leq n. \tag{5.2}$$

In particular, we can see that the ith column is a linear combination (only) of the orthonormal vectors $w^1, \ldots, w^i$, $1 \leq i \leq n$. Note that $||u^i||$, which is the coefficient of w^i in (5.2), is strictly positive.

Denote by $Q \in R^{m \times n}$ the matrix whose columns are these w^i, $1 \leq i \leq n$, vectors. From (5.2) we can see that

$$A = \begin{pmatrix} \vdots & \vdots & \cdots & \vdots \\ v^1 & v^2 & \cdots & v^n \\ \vdots & \vdots & \cdots & \vdots \end{pmatrix} = \begin{pmatrix} \vdots & \vdots & \cdots & \vdots \\ w^1 & w^2 & \cdots & w^n \\ \vdots & \vdots & \cdots & \vdots \end{pmatrix} \begin{pmatrix} R_{11} & R_{12} & cdots & R_{1n} \\ 0 & R_{22} & \cdots & R_{2n} \\ \vdots & \vdots & \vdots & \vdots \\ 0 & 0 & \cdots & R_{nn} \end{pmatrix} = QR,$$

for some upper-triangular matrix $R \in R^{n \times n}$. Note that the entries which are in the diagonal of R or above it are $R_{ii} = ||u^i||$, $1 \leq i \leq n$ and $R_{ij} = (v^i)^t w^j$, $1 \leq i < j \leq n$. In particular, the diagonal entries are all strictly positive. Expressing a full-rank tall matrix $A \in R^{m \times n}$, whose rank equals m, as a product between two matrices: $Q \in R^{m \times n}$, which comes with orthonormal columns, and a square upper-triangular matrix $R \in R^{n \times n}$ with a positive diagonal, is called the *QR factorization* of A. Note that the factorization process when applied to A, is basically applying the Gram–Schmidt procedure with the columns of A as input. Also note that if A is not a full-rank matrix then some of the columns of Q turned out to be all zeros, and then any choice for the corresponding column in R will do. Moreover, the number of non-zero columns in Q coincides with $rank(A)$.

Computing R given A and Q is quite straightforward. Specifically, it can be easily verified that $Q^t Q = I$, as the columns of Q are orthonormal. Thus, multiplying both sides of the equality $A = QR$ by Q^t from the left, implies that $Q^t A = Q^t Q R = (Q^t Q)R = IR = R$. We therefore conclude that $R = Q^t A$. This result is useful for the cases where A and Q are given and one looks for R.

Example 1. Consider the example in page 35. In this case we have the following QR factorization:

$$A = \begin{pmatrix} 1 & 0 & 0 \\ 1 & 1 & 0 \\ 1 & 1 & 1 \end{pmatrix} = \begin{pmatrix} \frac{1}{\sqrt{3}} & \frac{-2}{\sqrt{6}} & 0 \\ \frac{1}{\sqrt{3}} & \frac{1}{\sqrt{6}} & \frac{-1}{\sqrt{2}} \\ \frac{1}{\sqrt{3}} & \frac{1}{\sqrt{6}} & \frac{1}{\sqrt{2}} \end{pmatrix} \begin{pmatrix} \sqrt{3} & \frac{2}{\sqrt{3}} & \frac{1}{\sqrt{3}} \\ 0 & \frac{\sqrt{6}}{3} & \frac{1}{\sqrt{6}} \\ 0 & 0 & \frac{1}{\sqrt{2}} \end{pmatrix}.$$

Example 2. The following matrix, as we will show below, plays an important role in data science. For a given vector $x \in R^n$, let

$$A = \begin{pmatrix} 1 & x_1 \\ 1 & x_2 \\ \vdots & \vdots \\ 1 & x_m \end{pmatrix}.$$

Then, its QR-factorization is

$$Q = \frac{1}{\sqrt{m}} \begin{pmatrix} 1 & \frac{x_1 - \bar{x}}{SD(x)} \\ 1 & \frac{x_2 - \bar{x}}{SD(x)} \\ \vdots & \vdots \\ 1 & \frac{x_m - \bar{x}}{SD(x)} \end{pmatrix} \quad \text{and} \quad R = \sqrt{m} \begin{pmatrix} 1 & \bar{x} \\ 0 & SD(x) \end{pmatrix}. \tag{5.3}$$

Exercise 5.8

Let

$$A = \begin{pmatrix} 2 & 0 & 0 \\ 0 & 4 & 1 \\ 0 & 0 & 9 \\ 1 & 7 & 0 \\ 0 & 0 & -3 \end{pmatrix}.$$

(a) Perform the Gram–Schmidt process on the columns of A.
(b) Find the QR factorization of A.

Exercise 5.9

Compute the QR factorization of the following matrix:

$$A = \begin{pmatrix} 12 & -51 & 4 \\ 6 & 167 & -68 \\ -4 & 24 & -41 \end{pmatrix}.$$

5.7 Row and Column Operations

Let $A \in R^{m \times n}$ be a matrix. We next define three operations which can be formed on the rows of A. Specifically, we state each one of them in words, coupled with defining a square matrix $E \in R^{m \times m}$ such that the product EA yields the same matrix as applying the corresponding row operation to A. Such a matrix E is called a *row-operation matrix*.

1. Multiplying all entries of the ith row by a constant $c \neq 0$. In this case the non-zero entries in E are $E_{jj} = 1$, $1 \leq j \neq i \leq n$, and $E_{ii} = c$.
2. Swap between the ith and the jth row. In this case the non-zero entries of E are $E_{kk} = 1$, $k \neq i, j$, and $E_{ij} = E_{ji} = 1$.
3. Subtracting c times the jth row from the ith row. In this case the non-zero entries of E are $E_{kk} = 1$, $1 \leq k \leq n$, and $E_{ij} = -c$.

Note that in all above three cases E is in fact the unit matrix where the corresponding row operation already.

Observe that if E is a row operation matrix, then the same is the case with E^t. Specifically, for the first and second row operations, $E^t = E$, while for the third, E^t swaps the role of the two rows i and j. Note also that any row operation can be reversed by applying some other (but related) row operation. Note that in the second operation defined above, this in fact is done by the repeating of the same operation. We leave it to the reader to find what are the matrices in the two other cases.

Let $E \in R^{n \times n}$ be a row operation matrix. Consider the matrix AE. It is easy to see that the corresponding operation was performed but on the columns of A. Specifically, since $AE = (E^t A^t)^t$, we can say that right multiplying A by E is equivalent to performing the row operation E^t on A^t and then taking the transpose of the resulting matrix.

Theorem

Theorem 5.8. *A row (or column) operation preserves the rank of a matrix.*

Proof. It is possible to see that any vector which is in the span of the rows of A is also in the span of the rows of the matrix which is derived from A by any row operation. Since the same logic is applied when the row operation is reserved, our proof is complete. □

5.8   Echelon Matrices and the Rank of a Matrix

Definition

Definition 5.2. Let $A \in R^{m \times n}$. For $1 \leq i \leq n$, let $k(i)$ with $0 \leq k(i) \leq n$, be the highest indexed column of A with $A_{ij} = 0$ for all j, $0 \leq j \leq k(i)$. Then, the matrix A is said to be an echelon matrix if for any i, $1 \leq i \leq n$, with $k(i) < n$, then $d(i+1) > d(i)$. Otherwise, namely if $d(i) = n$, then $d(i+1) = n$.

An echelon matrix is in fact a matrix in which, maybe with the exception of the first row, all rows commence with a series of zeros. Moreover, each of these series is longer by at least one from the previous series, unless of course the previous one is already a zero row, making all the following zero rows too. It is certainly a possibility that the last row, or some of the last rows, are full of zeroes. In fact, there is much interest in this case. An example for 4×6 echelon matrix is:

$$\begin{pmatrix} 3 & 2 & 1 & -1 & 2 & 8 \\ 0 & 1 & -2 & 0 & 3 & 4 \\ 0 & 0 & 0 & -1 & 2 & 5 \\ 0 & 0 & 0 & 0 & 0 & 0 \end{pmatrix}. \tag{5.4}$$

Theorem

Theorem 5.9. *All non-zero rows in an echelon matrix are linearly independent. Hence, the rank of an echelon matrix equals the number of its non-zero rows. In particular, its non-zero rows form a basis for the linear subspace span by its rows. Finally, the columns which correspond to the echelons, namely those in which the first in their row non-zero entry appears, form a basis for the linear subspace spanned by its columns.*

Proof. Express the row vector $\underline{0} \in R^{1 \times n}$ as a linear combination of the non-zero rows. Observe that due to the structure of the echelon matrix, the coefficient of the first row needs to be zero. Once we know that, we can see that the same is the case with the coefficient of the second row, etc. The same argument goes with respect to the selected columns where now we go from the right column to the left. □

Note that the rank of the matrix in (5.4) is three. Its first three rows form a basis for the row subspace while columns 1, 2 and 4 form a basis for the column subspace. Combining all stated so far in this section, we can see that if within a finite number of row operations we convert a given matrix into an echelon matrix, we will learn its rank as it equals the number of non-zero rows. What is left to see is that this can be done for any matrix. This is indeed the case and in fact, the procedure we design below also derives a basis for the linear subspace spanned by the rows of A. Based on Theorem 5.9 a by-product will be a basis for the subspace spanned by the columns. Before we do that we remind the reader that another procedure which does exactly that is the Gram–Schmidt process.

The goal here is to find the rank of a matrix. Thus, any performed row operations do not change the rank of the matrix. The procedure we described next does exactly that but it terminates with an echelon matrix, whose rank is easily read using Theorem 5.9.

Row operations leading to an echelon matrix

Let $A \in R^{m \times n}$. For convenience assume A to be wide, namely $m \leq n$. Otherwise, deal with A^t instead. Let j be the first non-zero column of A. Assume that $A_{1j} \neq 0$. Otherwise swap the first row with any other row with $A_{ij} \neq 0$. Also, assume $A_{1j} = 1$. Otherwise, divide all entries in row i

by A_{1j}. Then, for any row i, $2 \le i \le m$, subtract A_{ij} times the first row from it, thereby zeroing the A_{ij} entry.[a] Repeat this when considering the resulting matrix whose first row is row 2 and its first column is column $j+1$, etc. Note that this procedure terminates after at most m steps of moving from one row to next. The procedure can terminate after less than m steps in the case where the submatrix left is full of zeros. By construction, the final matrix is an echelon matrix. As said, the number of its non-zero rows is the rank of the matrix we are looking for.

Example. $j = 1$.

$$A = \begin{pmatrix} 1 & 2 & -3 \\ 2 & 6 & -11 \\ 1 & -2 & 7 \end{pmatrix}$$

$$R_3 \leftarrow R_3 - R_1$$

$$\begin{pmatrix} 1 & 2 & -3 \\ 2 & 6 & -11 \\ 0 & -4 & 10 \end{pmatrix}$$

$$R_2 \leftarrow R_2 - 2R_1$$

$$\begin{pmatrix} 1 & 2 & -3 \\ 0 & 2 & -5 \\ 0 & -4 & 10 \end{pmatrix}.$$

$j = 2$.

$$R_3 \leftarrow R_3 + 2R_2$$

$$\begin{pmatrix} 1 & 2 & -3 \\ 0 & 2 & -5 \\ 0 & 0 & 0 \end{pmatrix}.$$

We have reached an echelon matrix whose rank is two. Note that we ended up here with an upper-triangular matrix. This is not a coincidence: Converting a square matrix to an echelon matrix, results in an upper-triangular matrix.

---

[a] Note that this operation is superfluous in the case where $A_{ij} = 0$.

Chapter 6

Invertible Matrices and the Inverse Matrix

6.1 Left-Inverses[a]

Let $A \in R^{m \times n}$. The matrix X is said to be a *left-inverse* of A if

$$XA = I.$$

Note that $X \in R^{n \times m}$ (as the dimension of A^t) and $I \in R^{n \times n}$. In the case where such an X exists, we say that A is *left-invertible*. Note that if X is a left-inverse of A, then not only that XA is well defined, but also AX is well defined. However, by no means we claim that AX equals $I \in R^{m \times m}$.

Examples.

- The scalar a is left-invertible if and only if $a \neq 0$. In which case there exists a unique left-inverse: $1/a$.
- For a vector $v \in R^{n \times 1}$, and for some i, $1 \leq i \leq n$, if $v_i \neq 0$, then $x \in R^{1 \times n}$ with $x = \frac{1}{v_i} e_i^t$, $1 \leq i \leq n$, is a left-inverse of v.
- If A is a matrix with orthonormal columns, then $A^t A = I$. In other words, A is left-invertible and A^t is a left-inverse of A.
- If X_1 and X_2 are left-inverses of A, then the same is the case with $\alpha X_1 + \beta X_2$ for any scalars α and β with $\alpha + \beta = 1$.

Theorem

Theorem 6.1. *If $A \in R^{m \times n}$ is left-invertible then its columns are linearly independent. In particular, $m \geq n$, namely A is a square or a tall matrix.*

---

[a]The first three sections in this chapter are based on [3].

Proof. Aiming for a contradiction, assume that a vector $x \neq \underline{0}$ with $Ax = \underline{0}$ exists. Let X be a left-inverse of A. Then

$$\underline{0} = X(Ax) = (XA)x = Ix = x,$$

which is a contradiction. $\qquad\square$

The converse of this theorem is also true. This issue is dealt with later in the second item of Theorem 7.1.

Consider the linear system of equations $Ax = b$. We know from previous sections that a solution exists if and only if b is in the linear subspace spanned by the columns of A. Moreover, when such a solution exists, it is unique if and only if the columns of A are linearly independent. This leads to the following theorem.

Theorem

Theorem 6.2. *Let A be a matrix and assume it has a left-inverse X. Then $Ax = b$ has a solution x if and only if $x = Xb$. In particular, Xb is the unique solution.*

Proof. Assume there exists an x such that $Ax = b$. Since A possesses a left-inverse, we know from the previous theorem that the columns of A are linearly independent and hence this solution, x, is unique. We next show that $x = Xb$. Indeed,

$$Xb = X(Ax) = (XA)x = Ix = x. \qquad\square$$

Note that the existence of a left-inverse X allows us to learn that for a given b whether or not a vector x with $Ax = b$ exists (or, equivalently, if b lies in the linear subspace spanned by the columns of A): All one needs to do is to check whether or not AXb equals b. Note in passing that the product AX is well-defined.

Note that if $m > n$, namely there are more equations than unknowns, the system is said to be *over-determined*. Loosely speaking, in most cases there is no solution to the system $Ax = b$. See Example 1 in Section 2.4.

What this section certainly lacks is an algorithm or a formula for computing a left-inverse when it exists or identifying that A is not left-invertible. This issue will be dealt with at length in Chapter 7 which in fact is devoted for that, although the point of departure there is different.

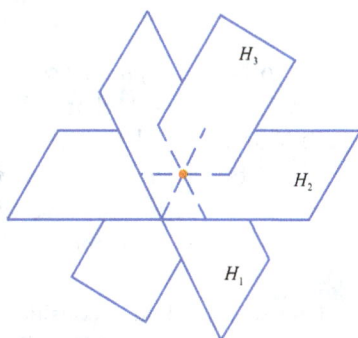

Unique Solution

Figure 6.1. Illustration of a unique solution.

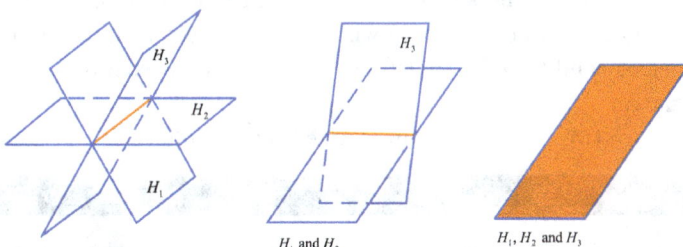

H_1 and H_2 $\qquad$ H_1, H_2 and H_3

Infinite Solutions

Figure 6.2. A case of infinite number of solutions.

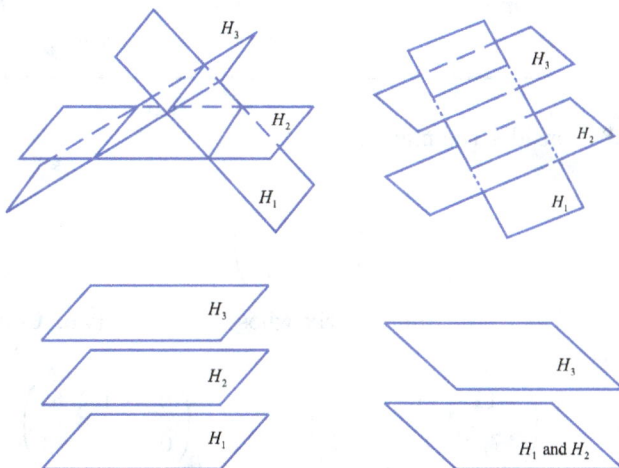

No Solution

Figure 6.3. A case of no solution.

6.2   Right-Inverses

A matrix $A \in R^{m \times n}$ is said to have a *right-inverse* if A^t has a left-inverse. In other words, if a matrix X exists such that $AX = I$ (and then X^t is a left-inverse of A^t). Note that if $A \in R^{m \times n}$ then $I \in R^{m \times m}$ and $X \in R^{n \times m}$. If such a matrix X exists, we say that A is *right-invertible*. Based on Theorem 6.1, we can say that if A is right-invertible, then its rows are linearly independent. Moreover, $m \leq n$, namely A is a square or a wide matrix.

Suppose X is a right-inverse of A and consider the set of equations $Ax = b$. It is easy to see that $A(Xb) = (AX)b = Ib = b$. In other words, the existence of a right-inverse X implies the existence of a solution x for any right-hand side b. Moreover, it states one: Xb. This, of course, says nothing on the uniqueness of the solution and indeed more conditions are needed to make it unique. As we will see in the next chapter, a necessary and sufficient condition for that (once a right-inverse exists) is that A is a square matrix.

The above two results are summarized by the following theorem.

Theorem

Theorem 6.3. *If a matrix $A \in R^{n \times n}$ possesses a right-inverse X, then*

- *its rows are linearly independent,*
- *for any vector $b \in R^n$ there exists a solution to the system of equations $Ax = b$. For example, Xb is such a solution.*

Example.[b] Consider the matrix

$$A = \begin{pmatrix} -3 & -4 \\ 4 & 6 \\ 1 & 1 \end{pmatrix}.$$

Note that A is a full rank matrix whose degree is two. Consider two more matrices:

$$B = \frac{1}{9}\begin{pmatrix} -11 & -10 & 16 \\ 7 & 8 & -11 \end{pmatrix} \qquad C = \frac{1}{2}\begin{pmatrix} 0 & -1 & 6 \\ 0 & 1 & -4 \end{pmatrix}.$$

---

[b]This example appears in [3].

It is possible to check that the matrices B and C satisfy $BA = CA = I$, namely they are both left-inverses of A. Consider the set of equation $Ax = b$ with $b^t = (1, -2, 0)$. Check that $Bb = Cb = (1, -1)^t$ and that $ABb = b$. In particular, b is in the linear subspace spanned by the two columns of A. Now change b to $b^t = (1, -1, 0)$ and notice that $Cb = (0.5, -0.5)$, but now $ACb \neq b$. Hence, the set $Ax = b$ does not have a solution for this vector b. In particular, b is not in the linear subspace spanned by the columns of A. Since B and C are left-inverses of A, then B^t and C^t are right-inverses for A^t. Hence, for any b, both $B^t b$ and $C^t b$ are solutions to $A^t x = b$. For example, if $b^t = (1, 2)$, they equal $(1/3, 2/3, -2/3)$ and $(0, 1/2, -1)$, respectively. Note that these two vectors are different.

Exercise 6.1

A set C is convex if for any pair of points $x, y \in C$ and for every $\lambda \in [0, 1]$ it holds that $\lambda x + (1 - \lambda)y \in C$. Now, let X, Y be two left-inverse matrices of A. Show that any convex combination of these matrices yields a left-inverse of A. Conclude that if more than one left-inverse of A exists, then there is an infinite number of left-inverses of A.

Exercise 6.2

Let:

$$A = \begin{pmatrix} -3 & 4 & 1 \\ -4 & 6 & 1 \end{pmatrix}, \quad B = \frac{1}{9} \begin{pmatrix} -11 & 7 \\ -10 & 8 \\ 16 & -11 \end{pmatrix}, \quad C = \frac{1}{2} \begin{pmatrix} 0 & 0 \\ -1 & 1 \\ 6 & -4 \end{pmatrix}, \quad b = \begin{pmatrix} 1 \\ -2 \end{pmatrix}$$

(a) Show that the matrices B and C are right-inverses of A. Express by B and C solutions to the set of linear equations $Ax = b$.

(b) Show that the difference between the above solutions yields a solution to the homogeneous set of equations $Ax = 0$.

6.3 Invertible Matrices

Suppose the matrix $A \in R^{m \times n}$ has a left-inverse X (and hence $m \geq n$) and a right-inverse Y (and hence $m \leq n$). Thus, the first thing to observe is that A needs to be a square matrix. Moreover

$$Y = (XA)Y = X(AY) = X.$$

On top of the obvious conclusion that $Y = X$, we get that this inverse matrix is unique: indeed, had there been another left inverse, say X', we would still get that $Y = X'$ and then $X' = X$. In this case we call this unique left and right-inverse the *inverse* of A. In this case A is said to be *invertible* or *regular*. Otherwise, it is called *not-invertible* or *singular*. The inverse of A (when exists) is denoted by A^{-1} and it satisfies

$$A^{-1}A = AA^{-1} = I. \tag{6.1}$$

Note that A is now the inverse of A^{-1} and in particular, A^{-1} is regular too. In other words, $(A^{-1})^{-1} = A$.

Remark. The reason behind the terminology of "the inverse of A" appears in (6.1). In particular, looking at Ax and $A^{-1}x$ as linear functions, we can see that each one of them is the inverse of the other: applying them consecutively (in both possible orders) is equivalent to applying the identity function, Ix.

Recall that the existence of the inverse here was based on the prior existence of both left and right-inverses. What is further true is the fact that if a square matrix A possesses a left (right, respectively) inverse then it also possesses a right (left, respectively) inverse. In particular, it is invertible. This is stated and proved in the following theorem.

Theorem

Theorem 6.4. *If a square matrix A has a left-inverse, then it also has a right-inverse. In particular, it is invertible.*

Proof. From Theorem 6.1 we learn that since $A \in R^{n \times n}$ has a left-inverse, its columns are linearly independent. Hence, the system of equations $Ax = b$ has a unique solution for any vector b. Let x_j be the solution to $Ax = e_j$, $1 \leq j \leq n$. Let $X \in R^{n \times n}$ be the matrix whose jth column is x_j, $1 \leq j \leq n$. It is easy to see that $AX = I$. Hence, A has a right-inverse.

Once a matrix A is known to be invertible, the system of equations $Ax = b$ has a unique solution: $x = A^{-1}b$. Indeed, this was expected in line with Section 6.2 as A^{-1} is a right-inverse of A. Note that x is a linear function of b and that A^{-1} defines the sensitivity of the solution to changes in the vector b: The derivative of x_i with respect to b_j equals A_{ij}^{-1}. This of course does leave open the computation question: Given A, how to derive

A^{-1} when it exists or to show that such an inverse does not exist when this is the case. This question is deferred to at a later section.

The following theorem summarizes what we have established by now.

Theorem

Theorem 6.5. *For a square matrix A the following statements are equivalent, namely they all hold if and only if one of them holds:*

1. *A has a right-inverse,*
2. *all columns of A are linearly independent,*
3. *all rows of A are linearly independent,*
4. *for any vector b, Ax = b possesses a solution,*
5. *for any vector b, Ax = b possesses a unique solution,*
6. *A has a left-inverse,*
7. *A is invertible.*

Proof. The proof follows from Theorems 6.1, 6.3 and 6.4, coupled with the fact that the number of linearly independent rows and the number of linearly independent columns, coincide (see Theorem 5.2). □

We next state an interesting theorem. We leave the proof to the reader as an exercise.

Theorem

Theorem 6.6. *Given a square matrix A, the system of equations Ax = b has a solution for any b if and only if Ax = b has a unique solution for a particular b.*

Exercise 6.3

Prove Theorem 6.6.

6.4   Solving a Set of Linear Equations

In the previous section, we dealt with the issue of the existence of the inverse of a square matrix. We defer the issue of computing this matrix to the next section. In the meantime consider the square set of linear equations $Ax = b$, where $A \in R^{n \times n}$ and $b \in R^n$. In the case where A^{-1} exists, we can see, by multiplying both hand sides of $Ax = b$ by A^{-1}, that $x = A^{-1}b$ is the unique solution.

What can we say in the case where A is not invertible? In this case let $m = rank(A)$ where $m < n$. The first thing to observe is that the set of equations $Ay = \underline{0}$ has a non-zero solution as the columns of A are not linearly independent. In fact, a non-trivial linear subspace of such solutions exists. The set of equations $Ay = \underline{0}$ is called the *homogeneous* equations associated with the set of equations $Ax = b$. Next observe that for any such solution y, if $Ax = b$, then $A(x + y) = b$, namely $x + y$ is also a solution to the original set of equations. Conversely, if $Ax^1 = Ax^2 = b$, that $c(x^2 - x^1)$ solves the homogeneous system for any constant c. From here we logically conclude that the system $Ax = b$ either does not have any solution at all or that it has an infinite number of solutions. In particular, for any $b \in R^n$, the system $Ax = b$ does not possess a unique solution.

Still in the non-invertible case, the next question is: for a given $b \in R^n$ what is the situation: are there infinite number of solutions or no solution at all? We next answer this question but wish to say first that we seldom encounter the first option. We will be more specific on that shortly. Without loss of generality assume that the first m rows of A are linearly independent. If this is not the case, just change the indices of the rows to make it like that. Denote the submatrix of A composed of these m rows by $C \in R^{m \times n}$. Note that the same change of indices is required with the entries of b (but not of x). Also, denote the first m entries of b by b_C. Then the first m equations out of the n in $Ax = b$, can be written as $Cx = b_C$.

Since C is a full-rank matrix, it comes with m (out of n) linearly independent columns. Assume without loss of generality that these are the first m columns of C. Otherwise, change the indices of the columns to make it this way. This time you need to make the same change in the entries of x (but not in b). Write $C = (B, N)$ where $B \in R^{m \times m}$ are these m linearly independent columns and $N \in R^{m \times (n-m)}$ comes with the rest of the columns of A.[c] Note that B is a full-rank $m \times m$ matrix. In particular, B^{-1} exists. Partition $x \in R^n$ in a similar fashion to (x_B, x_N), where $x_B \in R^m$ and $x_N \in R^{n-m}$. The set of equations $Cx = b_C$ can be written

---

[c]There is no consistency in the literature regarding terminology. Sometimes the first m variables are referred to as basic or pivotal variables, while the others $n - m$ are referred to as non-basic or free variables.

as $(B, N) \begin{pmatrix} x_B \\ x_N \end{pmatrix} = b_C$, or,

$$Bx_B + Nx_N = b_C.$$

Thus, $x_B = B^{-1}(b_C - Nx_N)$. In other words, one can select any x_N one wishes in order to solve the system $Cx = b_C$. Then, as long as $x_B = B^{-1}(b_C - Nx_N)$, x obeys $Cx = b_C$. Note that we have much freedom in selecting x_N. In fact, our choice of vectors for x_N in a linear subspace with a dimension of $n - m$.

Return now to the original set of equations $Ax = b$. Denote by A_i the ith row of A, $1 \leq i \leq n$. Consider the $m+1$th equation (which has been ignored so far). The $m + 1$th row of A can be expressed as a linear combination of the first m rows. Thus, for some α_i, $1 \leq i \leq m$, $A_{m+1} = \Sigma_{i=1}^{m}\alpha_i A_i$. Hence, for any solution to $Cx = b_C$, $A_{m+1}x = \Sigma_{i=1}^{m}\alpha_i A_i x = \Sigma_{i=1}^{m}\alpha_i b_i$. Thus, in order for $A_{m+1}x = b_{m+1}$ to hold too, which is a necessary condition for x to solve $Ax = b$, it is required that $b_{m+1} = \Sigma_{i=1}^{m}\alpha_i b_i$. Otherwise, which is usually the case, the set $Ax = b$ does not have a solution. Of course, a similar requirement applies for any b_{m+j}, $1 \leq j \leq n - m$. In particular, all these $n - m$ conditions must hold in order to have a solution to $Ax = b$. If this is the case, then as said on the unrestricted choice for x_N, there exist an infinite number of solutions to the set $Ax = b$.

Remark. Although above we assumed that the matrix A is a square matrix, this assumption can be easily removed and all the analysis holds for any wide or tall matrix.

Remark. The above comes with a few steps which at this stage seem theoretical as no algorithm is stated. Yet, this gap can be overcome. First, finding the set of linearly independent rows can be done via reducing the matrix to its echelon form as described in Section 5.8. Note that you need to perform the same operations on the right-hand side vector b in order to maintain the condition $Ax = b$. Observe that if at the end one of the zero rows comes with a non-zero entry in the right-hand side, no solution exists. Second, the matrix B can be composed out of the linearly independent columns found inspected at the echelon matrix. See Theorem 5.9. Finally, see the following section for how to invert an invertible matrix, leading to the all-important B^{-1} defined above.

Remark. In the special case where $b = \underline{0}$, the system of equations $Ax = \underline{0}$, is called *homogeneous*. They were introduced first above in Example 4 in

Section 2.2. In particular, it was argued there that the corresponding solution set is a linear subspace. Running through all possible unit vectors for x_N, we can generate $n - m$ solutions which span this linear subspace. In particular, its dimension is at most $n - m$. Yet, invoking Theorem 4.3, the dimension equals in fact to $n - m$.

6.5 Inverting Matrices

We start with a number of examples for the inverse of special matrices.

Examples.

- $I^{-1} = I$.
- If $A \in R^{n \times n}$ is a diagonal matrix with no zeros in the diagonal, then A^{-1} exists, is diagonal too and $A_{ii}^{-1} = 1/A_{ii}$, $1 \le i \le n$. Note that the requirement that all diagonal entries in a diagonal matrix should be non-zero is also a necessary condition for the existence of its inverse. It is possible in fact to see that the rank of this matrix coincides with the number of non-zero entries.
- Suppose $n = 2$.

$$A = \begin{pmatrix} a & b \\ c & d \end{pmatrix}.$$

Note that A is invertible if and only if $ad - bc \ne 0$: it is possible to see that otherwise each row (column, respectively) is a scalar multiplication of the other row (column, respectively), making them not linearly independent. Then,

$$A^{-1} = \frac{1}{ad - bc} \begin{pmatrix} d & -b \\ -c & a \end{pmatrix}.$$

- The columns of A are orthonormal if and only if $A^t = A^{-1}$. Equivalently, if $A^t A = I = AA^t$. A matrix with property is called a *unitary* matrix. The idea behind the terminology of unitary is the easy to see fact that for a unitary matrix U, $\|Ux\| = \|x\|$. Thus the operation of multiplying a vector by a unitary matrix preserves its magnitude, as summarized by its norm. In particular, only the direction of x is changed (or rotated) from $x/\|x\|$ to $Ux/\|x\|$. Note that a unitary matrix also preserves the distance between two vectors: $\|Uv - Uu\| = \|U(v - u)\| = \|v - u\|$.

We can collect a few more results related to invertible matrices.

- $(A^t)^{-1} = (A^{-1})^t$.
- $(A^k)^{-1} = (A^{-1})^k$ for any positive or negative k.
- If A and B (of the same dimension) are invertible then AB is invertible too. Moreover

$$(AB)^{-1} = B^{-1}A^{-1}.$$

Proof.

$$(B^{-1}A^{-1})(AB) = B^{-1}(A^{-1}A)B = B^{-1}IB = B^{-1}B = I. \qquad \square$$

- A row operation matrix is invertible. Moreover, the inverse of a row operation matrix, is a row operation matrix itself.
- If A is not invertible, then the same is the case with AB for any matrix B.

Proof. If B itself is not invertible, then there exists a vector $x \neq \underline{0}$ such that $Bx = \underline{0}$. Then, $ABx = \underline{0}$ too, implying that AB is not invertible. Since A is not invertible, there exists a vector $y \neq 0$ such that $Ay = \underline{0}$. If B is invertible, then there exists a vector $x \neq \underline{0}$ such that $y = Bx$ (it is $x = B^{-1}y$). Then $ABx = Ay = \underline{0}$, implying that AB is not invetible as required. $\qquad \square$

> **Theorem**
>
> **Theorem 6.7.** *An upper or lower triangular-matrix with non-zero diagonal entries is invertible.*

Proof. Suppose $A \in R^{n \times n}$ is lower-triangular and one looks for an $x \in R^n$ which solves $Ax = \underline{0}$. We need to show that $x = \underline{0}$ is the unique solution. This set of equations can be put as

$$\begin{pmatrix} A_{11} & 0 & 0 & \cdots & 0 \\ A_{21} & A_{22} & 0 & \cdots & 0 \\ A_{31} & A_{32} & A_{33} & \cdots & 0 \\ \vdots & \vdots & \vdots & \vdots & \vdots \\ A_{n1} & A_{n2} & A_{n3} & \cdots & A_{nn} \end{pmatrix} \begin{pmatrix} x_1 \\ x_2 \\ \vdots \\ \vdots \\ x_n \end{pmatrix} = \begin{pmatrix} 0 \\ 0 \\ \vdots \\ \vdots \\ 0 \end{pmatrix}. \qquad (6.2)$$

By forward substitutions we can argue that $x_1 = x_2 = \cdots = x_n = 0$ is the unique solution. Indeed, consider the first equation. Since $A_{11} \neq 0$, the unique solution is $x_1 = 0$. Use this value at the second equation and again since $A_{22} \neq 0$, we get $x_2 = 0$. Going on this way completes the proof. $\square$

A similar argument applies for an upper-triangular matrix. Alternatively, note that if A is upper-triangular then A^t is lower-triangular and the matrix is invertible if and only if its transpose is invertible. Note that had one of the diagonal entries been zero, the solution would not be unique and in particular, the matrix would not be invertible.

Consider again the set of equations stated in (6.2) and replace the right-hand side with some other vector. Then, solving this system can be done in the same way by forward substitutions. In particular, the resulting solution is unique. Inspect now the special case where the right-hand side vector is the unit vector e_j. Then, it is possible to see that the resulting solution can serve as the jth column of the inverse of A (if exists). Yet, since this can be done for any e_j, $1 \leq j \leq n$, the existence of the inverse is hence established. Denote the solution for $Ax = e_j$ by the vector whose entries are A_{ij}^{-1}, $1 \leq i \leq n$. It is possible to see that the first $j - 1$ entries turned out to be zeros, namely, $A_{ij}^{-1} = 0$, $1 \leq i < j \leq n$. Thus, we can conclude that if A is lower-triangular, then the same is the case with its inverse (when exists).

Example. For the matrix A which appears at page 60, we have already computed its QR factorization:

$$A = \begin{pmatrix} 1 & 0 & 0 \\ 1 & 1 & 0 \\ 1 & 1 & 1 \end{pmatrix} = \begin{pmatrix} \frac{1}{\sqrt{3}} & \frac{-2}{\sqrt{6}} & 0 \\ \frac{1}{\sqrt{3}} & \frac{1}{\sqrt{6}} & \frac{-1}{\sqrt{2}} \\ \frac{1}{\sqrt{3}} & \frac{1}{\sqrt{6}} & \frac{1}{\sqrt{2}} \end{pmatrix} \begin{pmatrix} \sqrt{3} & \frac{2}{\sqrt{3}} & \frac{1}{\sqrt{3}} \\ 0 & \frac{\sqrt{6}}{3} & \frac{1}{\sqrt{6}} \\ 0 & 0 & \frac{1}{\sqrt{2}} \end{pmatrix} = QR.$$

We are interested in inverting the upper-triangular matrix R. Towards this end, we look for the following three systems of equations:

$$\begin{pmatrix} \sqrt{3} & \frac{2}{\sqrt{3}} & \frac{1}{\sqrt{3}} \\ 0 & \frac{\sqrt{6}}{3} & \frac{1}{\sqrt{6}} \\ 0 & 0 & \frac{1}{\sqrt{2}} \end{pmatrix} \begin{pmatrix} x_1 \\ x_2 \\ x_3 \end{pmatrix} = \begin{pmatrix} 1 \\ 0 \\ 0 \end{pmatrix}, \quad \begin{pmatrix} \sqrt{3} & \frac{2}{\sqrt{3}} & \frac{1}{\sqrt{3}} \\ 0 & \frac{\sqrt{6}}{3} & \frac{1}{\sqrt{6}} \\ 0 & 0 & \frac{1}{\sqrt{2}} \end{pmatrix} \begin{pmatrix} x_1 \\ x_2 \\ x_3 \end{pmatrix} = \begin{pmatrix} 0 \\ 1 \\ 0 \end{pmatrix},$$

and

$$\begin{pmatrix} \sqrt{3} & \frac{2}{\sqrt{3}} & \frac{1}{\sqrt{3}} \\ 0 & \frac{\sqrt{6}}{3} & \frac{1}{\sqrt{6}} \\ 0 & 0 & \frac{1}{\sqrt{2}} \end{pmatrix} \begin{pmatrix} x_1 \\ x_2 \\ x_3 \end{pmatrix} = \begin{pmatrix} 0 \\ 0 \\ 1 \end{pmatrix}.$$

Each of the above solution contributes a column to R^{-1}, which turns out to be

$$R^{-1} = \begin{pmatrix} \frac{1}{\sqrt{3}} & -\frac{\sqrt{6}}{3} & 0 \\ 0 & \frac{3}{\sqrt{6}} & -\frac{1}{\sqrt{2}} \\ 0 & 0 & \sqrt{2} \end{pmatrix}.$$

Our final point here considers the QR factorization of an invertible matrix. It was shown in Section 5.5 that if the columns of A are linearly independent (which in the case of a square matrix is equivalent to its invertibility), we get that $A = QR$ where (i) Q is an orthonormal matrix with $Q^{-1} = Q^t$, (ii) R is upper-triangular, and (iii) all diagonal entries in R are non-zero (in fact, positive) and hence R^{-1} exists. Thus,

$$A^{-1} = (QR)^{-1} = R^{-1}Q^{-1} = R^{-1}Q^t. \tag{6.3}$$

Thus, once a square matrix is written in its QR factorization form, finding its inverse is boiled down to finding the inverse of an upper-triangular matrix. As was described in the previous paragraph, this can be done by backward substitutions. To summarize, we can invert an invertible matrix $A \in R^{n \times n}$ by taking the following steps: Firstly, find its QR factorization, basically by doing the Gram–Schmidt process stated in Section 5.5. Secondly, invert the upper-diagonal matrix R by applying the backward substitution procedure n times. Finally, compute $R^{-1}Q^t$. This is our first procedure for inverting matrices. There will be more.

Example (cont.) Back to the above example:

$$A^{-1} = R^{-1}Q^t = \begin{pmatrix} \frac{1}{\sqrt{3}} & -\frac{\sqrt{6}}{3} & 0 \\ 0 & \frac{3}{\sqrt{6}} & -\frac{1}{\sqrt{2}} \\ 0 & 0 & \sqrt{2} \end{pmatrix} \begin{pmatrix} \frac{1}{\sqrt{3}} & \frac{1}{\sqrt{3}} & \frac{1}{\sqrt{3}} \\ -\frac{2}{\sqrt{6}} & \frac{1}{\sqrt{6}} & \frac{1}{\sqrt{6}} \\ 0 & -\frac{1}{\sqrt{2}} & \frac{1}{\sqrt{2}} \end{pmatrix} = \begin{pmatrix} 1 & 0 & 0 \\ -1 & 1 & 0 \\ 0 & -1 & 1 \end{pmatrix}.$$

Exercise 6.4

Consider the following matrix:

$$A = \begin{pmatrix} 12 & -51 & 4 \\ 6 & 167 & -68 \\ -4 & 24 & -41 \end{pmatrix}.$$

Given its QR factorization as computed in exercise 5.9, compute its inverse: A^{-1}.

6.6 Inverting a Matrix by Row Operations

In Eq. (6.3), we showed how to invert a matrix using its QR factorization. The question we pose now is if it is possible to invert a matrix without computing its QR factors first. We can draft to our mission the row operations defined in Section 5.6 and their use to convert a matrix to an echelon matrix. Suppose this procedure is applied to a square matrix. By definition, this procedure terminates with an upper-triangular matrix. Thus, for some row operation matrices $E_1, \ldots, E_k$,

$$E_k E_{k-1} \cdots E_1 A = U,$$

where U is an upper-triangular matrix. Since by Theorem 5.8

$$rank(E_k E_{k-1} \cdots E_1 A) = rank(A),$$

we conclude that A is invertible if and only if the same is the case with U, namely if and only if all diagonal entries in U are non-zero.[d]

As for computing A^{-1} we now have two options. The first is to invert the upper-triangular matrix as suggested in the previous section. Indeed, it is easy to see that $U^{-1} E_k \cdots E_1$ is the inverse of A:

$$(U^{-1} E_k \cdots E_1)A = U^{-1}(E_k \cdots E_1 A) = U^{-1}U = I.$$

The second option is as follows. Suppose we find another set of l row operation matrices, $E_{k+1}, \ldots, E_l$ such that

$$E_l \cdots E_{k+1} U = I,$$

then A^{-1} exists and it equals $E_{k+l} \cdots E_{k+1} E_k E_{k-1} \cdots E_1$. Note that this is possible as described next. Firstly, by row operations, make all diagonal entries in U equal to one. This can be done by dividing all entries in row i by U_{ii}, $1 \le i \le n$. Call this matrix U' and note that it is an upper-triangular matrix with all its diagonal entries being equal to 1. Secondly, all entries above the diagonal in U' can become zero by the corresponding

---

[d]Inspecting how the conversion to an echelon matrix is done, the last entry in the diagonal, namely U_{nn}, tells it all: A is invertible if and only if $U_{nn} \ne 0$.

row operations (without changing any zero entries into non-zero). Indeed, in order to zero U_{ij} with $1 \leq i < j \leq n-1$, do the following row operation:

$$R_i \leftarrow R_i - U'_{ij}R_j, \quad i+1 \leq j \leq n, \quad 1 \leq i \leq n-1.$$

Note that we have a total of $n(n-1)/2$ row operations which can be done in any order. Yet, it is somewhat easier to apply them in a decreasing order of i as one will meet more zeros from R_j, $i < j < n$.

In practice, what is usually done in order to compute $E_{k+l} \cdots$ $E_{k+1}E_k E_{k-1} \cdots E_1$, which of course equals $E_{k+l} \cdots E_{k+1}E_k E_{k-1} \cdots E_1 I$, is to apply the same row operations preformed on A (on the way toward I) concurrently on I.

There was a practical question that was ignored above: how do we select the row operations? Or put differently, how are we guaranteed that through a finite sequence of row operations we indeed obtain A^{-1}? We start with inspecting A_{11}. In case it equals zero, we perform the row operation which is interchanging the first row with some row i with $A_{i1} \neq 0$, $2 \leq i \leq n$. Of course this is possible: had the first column been full of zeros the matrix would not be invertible. The trick now is to use the entry in position $(1,1)$ as a pivot for zeroing all entries in the first column:

$$R_i \leftarrow R_i - \frac{A_{i1}}{A_{11}}R_1, \quad 2 \leq i \leq n.$$

This takes care of the first column. Now repeat this for all entries below the diagonal of the second column, while leaving the first row as is. Again, check first if the diagonal term is zero or not. In case it is, interchange row 2 with one of the rows i, $2 \leq i \leq n$, which has a non-zero entry in the second column. Finding such an entry is guaranteed. Next, zero all entries in the second column below the diagonal using the $(2,2)$ entry as a pivot. Then you do the same for the third column, keeping the first two rows untouched, etc. This leads to an upper-triangular matrix, with no zero entries on the diagonal. Then, make all diagonal entries one, by dividing each row with its diagonal entry. Now zero the entries above the diagonal. This is done first to the $n-th$ column, then the $(n-1)$th column, etc. See the following example. Note the matrices on the right which correspond to performing the same row operations on matrices initialized as the identity matrix.

Example.

$$A = \begin{pmatrix} 1 & -3 & 0 \\ 2 & -2 & 2 \\ 3 & -4 & 3 \end{pmatrix}, \quad I = \begin{pmatrix} 1 & 0 & 0 \\ 0 & 1 & 0 \\ 0 & 0 & 1 \end{pmatrix}$$

$$R_3 \leftarrow R_3 - 3R_1$$

$$\begin{pmatrix} 1 & -3 & 0 \\ 2 & -2 & 2 \\ 0 & 5 & 3 \end{pmatrix} \quad \begin{pmatrix} 1 & 0 & 0 \\ 0 & 1 & 0 \\ -3 & 0 & 1 \end{pmatrix}$$

$$R_2 \leftarrow R_2 - 2R_1$$

$$\begin{pmatrix} 1 & -3 & 0 \\ 0 & 4 & 2 \\ 0 & 5 & 3 \end{pmatrix} \quad \begin{pmatrix} 1 & 0 & 0 \\ -2 & 1 & 0 \\ -3 & 0 & 1 \end{pmatrix}$$

$$R_3 \leftarrow R_3 - \frac{5}{4}R_2$$

$$\begin{pmatrix} 1 & -3 & 0 \\ 0 & 4 & 2 \\ 0 & 0 & 0.5 \end{pmatrix} \quad \begin{pmatrix} 1 & 0 & 0 \\ -2 & 1 & 0 \\ -0.5 & -1.25 & 1 \end{pmatrix}.$$

Note that we have reached an upper-triangular matrix. The procedure goes on as follows:

$$R_2 \leftarrow R_2 - 4R_3$$

$$\begin{pmatrix} 1 & -3 & 0 \\ 0 & 4 & 0 \\ 0 & 0 & 0.5 \end{pmatrix} \quad \begin{pmatrix} 1 & 0 & 0 \\ 0 & 6 & -4 \\ -0.5 & -1.25 & 1 \end{pmatrix}$$

$$R_1 \leftarrow R_1 + \frac{3}{4}R_2$$

$$\begin{pmatrix} 1 & 0 & 0 \\ 0 & 4 & 0 \\ 0 & 0 & 0.5 \end{pmatrix} \quad \begin{pmatrix} 1 & 9/2 & -3 \\ -4 & 6 & -3 \\ -0.5 & -1.25 & 1 \end{pmatrix}.$$

Finally,

$$R_3 \leftarrow 2R_3 \quad \text{and} \quad R_2 \leftarrow \frac{1}{4}R_2$$

lead to the identity matrix. In particular,

$$A^{-1} = \begin{pmatrix} 1 & 4.5 & -3 \\ 0 & 1.5 & -1 \\ -1 & -2.5 & 2 \end{pmatrix}.$$

Exercise 6.5

Let:

$$A = \begin{pmatrix} 0 & 1 & 2 \\ 1 & 0 & 3 \\ 4 & -3 & 8 \end{pmatrix}, \quad b = \begin{pmatrix} 1 \\ 1 \\ 1 \end{pmatrix}.$$

(a) Solve the system of equations $Ax = b$ by transforming to the system $Lx = b'$, where L is an upper-triangular matrix, using row operations on A.

(b) Solve the same system of linear equations ($Ax = b$) by transforming the system $Lx = b'$ to $Ix = b''$, using row operations on L.

(c) Find A^{-1} using row operations on A and the same operation on I.

Exercise 6.6

Show that if E is a row operation matrix then its inverse exists. Specifically,

1. if E multiplies all entries of row i by a $c \neq 0$, then E^{-1} does the same but with $1/c$
2. if E swaps two rows, then $E^{-1} = E$, and
3. if E subtracts c times row j from row i, then E^{-1} adds c times row j to row i.

Conclude that the inverse, as expected, cancels the originally performed row operation.

6.7   Change of Bases and Similar Matrices

When we refer to a vector $x \in R^n$, we in fact have in mind $\Sigma_{i=1}^n x_i e_i$. In particular, x_i is the ith coefficient when we express x as a linear combination of the unit vectors which form the elementary basis for R^n. Suppose we have another basis in mind and let $V \in R^{n \times n}$ be the matrix whose columns coincide with the vectors in this basis. A natural question is what are the coefficients of these vectors when one expresses x as a linear combination thereof. The answer is simple. Specifically, let $\alpha \in R^n$ be the vector of coefficients. Then, $x = V\alpha$ and hence $\alpha = V^{-1}x$. If one takes another basis whose matrix is U, $x = U\beta$ and hence $\beta = U^{-1}x$. In passing, note that since $V\alpha = U\beta$, we have that $\alpha = V^{-1}U\beta$ and that $\beta = U^{-1}V\alpha$.

Example. Consider the vector $x = (1, 2, 3)$. Expressing x as a linear combination of the vectors $(1, 1, 0), (1, 1, 2), (2, 0, 1)$, our task is to find the corresponding coefficients. Note that the three vectors form a linear basis of R^3 (a fact that will be established anyhow when we show that the matrix formed out of these vectors as its columns is invertible). The corresponding matrix V (whose columns form the basis) is

$$V = \begin{pmatrix} 1 & 1 & 2 \\ 1 & 1 & 0 \\ 0 & 2 & 1 \end{pmatrix}.$$

In order to find the coefficients, we need to invert this matrix. We do that through a sequence of row operations $E_l \cdots E_1$ which transform V into the identity matrix. The product $E_l \cdots E_1$ is therefore the inverse of V, which can also be obtained by applying the same row operations to the identity matrix. Details are given in the following table, describing all operations step by step on the original and identity matrices:

Row operation	Operation matrix	Result on original matrix	Result on identity matrix
$R_2 \leftarrow R_2 - R_1$	$E_1 = \begin{pmatrix} 1 & 0 & 0 \\ -1 & 1 & 0 \\ 0 & 0 & 1 \end{pmatrix}$	$\begin{pmatrix} 1 & 1 & 2 \\ 0 & 0 & -2 \\ 0 & 2 & 1 \end{pmatrix}$	$\begin{pmatrix} 1 & 0 & 0 \\ -1 & 1 & 0 \\ 0 & 0 & 1 \end{pmatrix}$
$R_2 \leftarrow R_2 + 2R_3$	$E_2 = \begin{pmatrix} 1 & 0 & 0 \\ 0 & 1 & 2 \\ 0 & 0 & 1 \end{pmatrix}$	$\begin{pmatrix} 1 & 1 & 2 \\ 0 & 4 & 0 \\ 0 & 2 & 1 \end{pmatrix}$	$\begin{pmatrix} 1 & 0 & 0 \\ -1 & 1 & 2 \\ 0 & 0 & 1 \end{pmatrix}$
$R_3 \leftarrow R_3 - \frac{1}{2}R_2$	$E_3 = \begin{pmatrix} 1 & 0 & 0 \\ 0 & 1 & 0 \\ 0 & -0.5 & 1 \end{pmatrix}$	$\begin{pmatrix} 1 & 1 & 2 \\ 0 & 4 & 0 \\ 0 & 0 & 1 \end{pmatrix}$	$\begin{pmatrix} 1 & 0 & 0 \\ -1 & 1 & 2 \\ 1/2 & -1/2 & 0 \end{pmatrix}$
$R_2 \leftarrow \frac{1}{4}R_2$	$E_4 = \begin{pmatrix} 1 & 0 & 0 \\ 0 & 0.25 & 0 \\ 0 & 0 & 1 \end{pmatrix}$	$\begin{pmatrix} 1 & 1 & 2 \\ 0 & 1 & 0 \\ 0 & 0 & 1 \end{pmatrix}$	$\begin{pmatrix} 1 & 0 & 0 \\ -1/4 & 1/4 & 1/2 \\ 1/2 & -1/2 & 0 \end{pmatrix}$
$R_1 \leftarrow R_1 - R_2$	$E_5 = \begin{pmatrix} 1 & -1 & 0 \\ 0 & 1 & 0 \\ 0 & 0 & 1 \end{pmatrix}$	$\begin{pmatrix} 1 & 0 & 2 \\ 0 & 1 & 0 \\ 0 & 0 & 1 \end{pmatrix}$	$\begin{pmatrix} 5/4 & -1/4 & -1/2 \\ -1/4 & 1/4 & 1/2 \\ 1/2 & -1/2 & 0 \end{pmatrix}$
$R_1 \leftarrow R_1 - 2R_3$	$E_6 = \begin{pmatrix} 1 & 0 & -2 \\ 0 & 1 & 0 \\ 0 & 0 & 1 \end{pmatrix}$	$\begin{pmatrix} 1 & 0 & 0 \\ 0 & 1 & 0 \\ 0 & 0 & 1 \end{pmatrix}$	$\begin{pmatrix} 1/4 & 3/4 & -1/2 \\ -1/4 & 1/4 & 1/2 \\ 1/2 & -1/2 & 0 \end{pmatrix}$

The inverse V^{-1} is given in the lower right box of the table. Standard computation shows that it coincides with $V^{-1} = E_6 \cdots E_1$. Thus,

$$V^{-1} = \begin{pmatrix} 1/4 & 3/4 & -1/2 \\ -1/4 & 1/4 & 1/2 \\ 1/2 & -1/2 & 0 \end{pmatrix}.$$

Finally, to find the coefficients we only need to compute $V^{-1}x$, which leads to:

$$V^{-1}x = \begin{pmatrix} 1/4 & 3/4 & -1/2 \\ -1/4 & 1/4 & 1/2 \\ 1/2 & -1/2 & 0 \end{pmatrix} \begin{pmatrix} 1 \\ 2 \\ 3 \end{pmatrix} = \begin{pmatrix} 1/4 \\ 7/4 \\ -1/2 \end{pmatrix}.$$

Let $A \in R^{m \times n}$ and consider the linear function $y = Ax$. Our default assumes that both x are y are written as linear combination of $\{e_i\}_{i=1}^n$ and $\{e_i\}_{i=1}^m$, respectively. Let $V \in R^{n \times n}$ and $U \in R^{m \times m}$ be basis matrices for R^n and R^m, respectively. We are now interested in the matrix of this linear function with respect to these two bases. Specifically, let $x = V\alpha$ and $y = U\beta$. Then,

$$Ax = AV\alpha = U\beta \Rightarrow \beta = U^{-1}AV\alpha,$$

So the answer is $U^{-1}AV$. Note that in the special case where $m = n$ and one uses the same basis for the domain and for the range, namely $U = V$, we get that the matrix we are after is $V^{-1}AV$.

Example. Let

$$A = \begin{pmatrix} 0 & 2 & 1 \\ 0 & -1 & -1 \\ 1 & 2 & 3 \end{pmatrix}.$$

Consider the columns of the following matrix:

$$V = \begin{pmatrix} 1 & -1 & 0 \\ 0 & 1 & -1 \\ 0 & 0 & 1 \end{pmatrix}.$$

They form a basis of R^3. What is the matrix that represents the same linear function as A but with respect to this basis? To answer this question we needed to compute V^{-1} which turns out to be

$$V^{-1} = \begin{pmatrix} 1 & 1 & 1 \\ 0 & 1 & 1 \\ 0 & 0 & 1 \end{pmatrix}.$$

Finally,

$$V^{-1}AV = \begin{pmatrix} 1 & 2 & 0 \\ 1 & 0 & 1 \\ 1 & 1 & 1 \end{pmatrix}$$

is the matrix we are after.

Definition

Definition 6.1. Two square matrices A and B are said to be *similar* if there exists a matrix V such that

$$B = V^{-1}AV.$$

An interpretation for this definition is as follows. Suppose a linear function Ax is given when the standard basis is used but one is interested in the matrix of this linear function when another basis, V, is assumed. What one needs to do is to first convert the given presentation into the one based on the standard basis. This is done by multiplying it with V. Then apply A to the result and finally multiply that by V^{-1} to convert the result back to the original basis.

Exercise 6.7

(a) Show that if A, B are similar then also $A - aI$ and $B - aI$ are similar for any $a \in R$.

(b) Let $A, B \in R^{n \times n}$ be similar matrices. Prove that $trace(A) = trace(B)$ where the *trace* of a square matrix is defined as the sum of the entries along its diagonal. Use the fact that $trace(CD) = trace(DC)$ for any square matrices C and D (both of the same dimension).

(c) Prove that the diagonal matrices

$$D_1 = \begin{pmatrix} a & 0 & 0 \\ 0 & b & 0 \\ 0 & 0 & c \end{pmatrix}, \quad D_2 = \begin{pmatrix} c & 0 & 0 \\ 0 & a & 0 \\ 0 & 0 & b \end{pmatrix},$$

are similar. What is the matrix S that satisfies $D_2 = SD_1S^{-1}$?

Exercise 6.8

Let $A = \begin{pmatrix} 1 & 2 \\ 3 & 4 \end{pmatrix}$ be the transformation matrix from R^2 to R^2 with respect to the standard basis. Let $\begin{pmatrix} 2 \\ 5 \end{pmatrix}$, $\begin{pmatrix} 1 \\ 3 \end{pmatrix}$ be another basis.

(a) Show that the transformation matrix with respect to that basis is $B = \begin{pmatrix} -5 & -8 \\ 6 & 10 \end{pmatrix}$.

(b) Check your result: let $S = \begin{pmatrix} 1 & 2 \\ 3 & 5 \end{pmatrix}$. Compute S^{-1} and verify that $A = SBS^{-1}$.

Exercise 6.9

Let $A \in R^{n \times n}$ be a square matrix. Let $E_1, \ldots, E_m$ and $F_1, \ldots, F_k$ two sequences of row operation matrices. Suppose that:

$$E_m \cdots E_1 A F_1 \cdots F_k = I.$$

(a) Explain why A is invertible.

(b) Express A^{-1} in terms of the above row operation matrices.

Exercise 6.10

Let $T : R^2 \to R^2$ be a linear transformation defined as:

$$T(x, y) = \begin{pmatrix} 3x + 4y \\ 2x - 5y \end{pmatrix}.$$

Given the following two bases:

$$E = \{e_1, e_2\}; \quad S = \left\{ \begin{pmatrix} 1 \\ 2 \end{pmatrix}, \begin{pmatrix} 2 \\ 3 \end{pmatrix} \right\}$$

(a) Find the transformation matrix A which represents T with respect to E, in both domain and range.

(b) Find the transformation matrix A which represents T with respect to S, in both domain and range.

(c) What is the transformation matrix if the domain subspace is expressed in terms of S and the range subspace is expressed in terms of $S =$
$$\left\{ \begin{pmatrix} -1 \\ 1 \end{pmatrix}, \begin{pmatrix} 2 \\ -1 \end{pmatrix} \right\}?$$

6.8 Special Inverses

6.8.1 Block matrices

> **Theorem**
>
> **Theorem 6.8.** *Let $A \in R^{n \times n}$. Its block presentation is*
> $$A = \begin{pmatrix} A_{11} & A_{21} \\ A_{21} & A_{22} \end{pmatrix},$$
> *where for some r, $1 \le r < n$, $A_{11} \in R^{r \times r}$, $A_{12} \in R^{r \times (n-r)}$, $A_{21} \in R^{(n-r) \times r}$ and $A_{22} \in R^{(n-r) \times (n-r)}$. Assume both A_{11} and A_{22} are invertible. Then,*
> $$A^{-1} = \begin{pmatrix} (A_{11} - A_{12} A_{22}^{-1} A_{21})^{-1} & -A_{11}^{-1} A_{12} (A_{22} - A_{21} A_{11}^{-1} A_{12})^{-1} \\ -A_{22}^{-1} A_{21} (A_{11} - A_{12} A_{22}^{-1} A_{21})^{-1} & (A_{22} - A_{21} A_{11}^{-1} A_{12})^{-1} \end{pmatrix}.$$

Proof. We look for matrices B_{11}, B_{12}, B_{21} and B_{22} such that
$$\begin{pmatrix} A_{11} & A_{21} \\ A_{21} & A_{22} \end{pmatrix} \begin{pmatrix} B_{11} & B_{21} \\ B_{21} & B_{22} \end{pmatrix} = \begin{pmatrix} I & \underline{0} \\ \underline{0} & I \end{pmatrix}.$$

The fact that $A_{21} B_{11} + A_{22} B_{21} = \underline{0}$, implies that $B_{21} = -A_{22}^{-1} A_{21} B_{11}$. Substituting that in the equation $A_{11} B_{11} + A_{21} B_{21} = I$ leads to $A_{11} B_{11} - A_{12} A_{22}^{-1} A_{21} B_{11} = I$, which can easily be solved for B_{11}, as required. This leads to the promised value for B_{21}. Solving for B_{21} and B_{22} is similar. $\square$

6.8.2 Rank-one updates

The following is the Sherman–Morrison formula. It says that if the inverse of a matrix is at hand, it is possible to find the inverse of a rank-one update of it more efficiently than by inverting the new matrix from scratch.

> **Theorem**
>
> **Theorem 6.9.** *Let $A \in R^{n \times n}$ be an invertible matrix. Let $v, u \in R^n$. The matrix $A + uv^t$ is invertible if and only if $v^t A^{-1} u \neq -1$, in which case*
>
> $$(A + uv^t)^{-1} = A^{-1} - \frac{1}{1 + v^t A^{-1} u} A^{-1} uv^t A^{-1}.$$
>
> *In particular, the correction of A^{-1} is a rank-one correction too.*

Proof. The result is established by a straightforward matrix multiplication, leading to the identity matrix. Indeed,

$$(A + uv^t) \left(A^{-1} - \frac{1}{1 + v^t A^{-1} u} A^{-1} uv^t A^{-1} \right)$$

$$= AA^{-1} + uv^t A^{-1} - \frac{1}{1 + v^t A^{-1} u} (uv^t A^{-1} + uv^t A^{-1} uv^t A^{-1})$$

$$= I + uv^t A^{-1} - \frac{1}{1 + v^t A^{-1} u} u(1 + v^t A^{-1} u) v^t A^{-1},$$

$$I + uv^t A^{-1} - uv^t A^{-1} - I,$$

as required. Clearly, the assumption that $v^t A^{-1} u \neq -1$ is necessary for the above to make sense. What is left is to argue that when $v^t A^{-1} u = -1$ the matrix $A + uv^t$ is not invertible. First, it implies that $u \neq \underline{0}$ but $(A + uv^t) A^{-1} u = Iu - u = \underline{0}$. Since $A^{-1} u \neq 0$, $A + uv^t$ is hence not invertible. $\square$

The above leads to an alternative procedure for computing the inverse of a matrix. It is based on the idea that any square matrix can be presented as the identity matrix plus the sum of n rank-one matrices. Then, its inverse can be derived by n consecutive applications of the Sherman–Morrison formula. Specifically, let A^j be the jth column of A and denote $A^j - e_j$ by u_j, $1 \leq j \leq n$. Recall that e_j is the jth unit vector, $1 \leq j \leq n$. Then, by replacing sequentially each of the (unit) columns of I by u_j, $1 \leq j \leq n$, we can see that

$$A = I + \sum_{j=1}^{n} u_j e_j^t.$$

The following algorithm is now immediate.

Input: an invertible matrix $A \in R^{n \times n}$.

Output: its inverse A^{-1}.

Set: $u_j = A^j - e_j$, $1 \le j \le n$.

Initialization: $A^{-1} = I$.

For $j = 1$ until $j = n$ do:

$$A^{-1} \leftarrow A^{-1} - \frac{1}{1 + e_j^t A^{-1} u_j} A^{-1} u_j e_j^t A^{-1}.$$

$$j \leftarrow j + 1.$$

Note that above $A^{-1}e_j$ ($e_j^t A^{-1}$, respectively) is the jth row (column, respectively) of the current A^{-1}, $1 \le j \le n$.

Example. Suppose

$$A = \begin{pmatrix} 1 & 1 \\ 1 & 2 \end{pmatrix} \quad \text{and hence} \quad u_1 = \begin{pmatrix} 0 \\ 1 \end{pmatrix}, \quad u_2 = \begin{pmatrix} 1 \\ 1 \end{pmatrix}.$$

Also,

$$A = \begin{pmatrix} 1 & 0 \\ 0 & 1 \end{pmatrix} + \begin{pmatrix} 0 & 0 \\ 1 & 0 \end{pmatrix} + \begin{pmatrix} 0 & 1 \\ 0 & 1 \end{pmatrix}.$$

First iteration:

$$A^{-1} \leftarrow \begin{pmatrix} 1 & 0 \\ 0 & 1 \end{pmatrix} - \frac{1}{1+0} \begin{pmatrix} 1 & 0 \\ 0 & 1 \end{pmatrix} \begin{pmatrix} 0 & 0 \\ 1 & 0 \end{pmatrix} \begin{pmatrix} 1 & 0 \\ 0 & 1 \end{pmatrix} = \begin{pmatrix} 1 & 0 \\ -1 & 1 \end{pmatrix}$$

Second (and final) iteration:

$$A^{-1} \leftarrow \begin{pmatrix} 1 & 0 \\ -1 & 1 \end{pmatrix} - \frac{1}{1+0} \begin{pmatrix} 1 & 0 \\ -1 & 1 \end{pmatrix} \begin{pmatrix} 0 & 1 \\ 0 & 1 \end{pmatrix} \begin{pmatrix} 1 & 0 \\ -1 & 1 \end{pmatrix} = \begin{pmatrix} 2 & -1 \\ -1 & 1 \end{pmatrix}$$

A hidden assumption above is that when called for, $1 + e_j^t A^{-1} u_j \ne 0$. There is no a-priori guarantee that this is the case, even when A^{-1} exists. For example, suppose

$$A = \begin{pmatrix} 0 & 1 \\ 1 & 0 \end{pmatrix}.$$

Clearly, $A^{-1} = A$. In particular, A^{-1} exists. Yet, at the first iteration, $1 + e_1^t I u_1 = 1 + e^t u_1 = 1 - 1 = 0$ and the algorithm will produce an error message. Note that $1 + e_2^t I u_2 = 0$, so swapping the order between the two rank-one updates does not lead to different results. Thus, a drawback of this procedure is that it fails to tell that its input matrix is not invertible even when this is the case. Put differently, if at some stage $1 + e_j^t A^{-1} u_j = 0$, one will not be able to tell the reason behind it.

Chapter 7

The Pseudo-Inverse Matrix, Projections and Regression

7.1 Least Squares Solutions[a]

We are now back to the issue of left-inverses. Our previous discussion of this subject dealt with some of the properties possessed by any left-inverse and we also state some sufficient conditions for their existence. Yet, we have not developed as of now any method for computing at least one left-inverse (when it exists). The exception was for the case of square matrices where the left-inverse coincides with the inverse (when it exists) which was dealt with in the previous chapter. Now that we are equipped with the concept of the inverses of square matrices, this gap on left-inverses for tall matrices can be filled up.

> ### Theorem
>
> **Theorem 7.1.** *Let $A \in R^{m \times n}$ be a tall matrix ($m \geq n$). Assume it is a full rank matrix (namely, $rank(A) = n$ and its columns are linearly independent). Then,*
>
> 1. *$A^t A$ (called the Gram matrix) is invertible,*
> 2. *$(A^t A)^{-1} A^t$ (called the Moore–Penrose (pseudo)-inverse) is a left-inverse of A. In particular, A has a left-inverse,*
> 3. *if a solution to the system of equations $Ax = b$ exists, then it is uniquely $(A^t A)^{-1} A^t b$.*

---

[a]This section is based on [3].

Proof. Aiming for a contradiction, assume there exists a non-zero $x \in R^n$ with $A^t A x = \underline{0}$. Then,

$$0 = x^t \underline{0} = x^t A^t A x = ||Ax||^2,$$

and hence $Ax = \underline{0}$, contradicting the fact that the columns of A are linearly independent. Also,

$$((A^t A)^{-1} A^t) A = (A^t A)^{-1} (A^t A) = I,$$

establishing the fact that $(A^t A)^{-1} A^t$ is a left-inverse of A. The final statement is now immediate from Theorem 6.2. □

Denote the matrix $(A^t A)^{-1} A^t$ by $A^\dagger$. It was just established that $A^\dagger A = I$, i.e., $A^\dagger$ is a left-inverse of A.

Example. Find the Moore–Penrose pseudo-inverse of the following matrix:

$$A = \begin{pmatrix} -3 & -4 \\ 4 & 6 \\ 1 & 1 \end{pmatrix}.$$

By simple calculations we obtain:

$$A^t A = \begin{pmatrix} 26 & 37 \\ 37 & 53 \end{pmatrix}.$$

The next step is to invert this matrix. Since the matrix is of dimension 2×2, we can use the inversion formula as given in page 74, and obtain:

$$(A^t A)^{-1} = \frac{1}{26 \cdot 53 - 37^2} \begin{pmatrix} 53 & -37 \\ -37 & 26 \end{pmatrix} = \frac{1}{9} \begin{pmatrix} 53 & -37 \\ -37 & 26 \end{pmatrix}$$

$$= \begin{pmatrix} 5.888889 & -4.111111 \\ -4.111111 & 2.888889 \end{pmatrix}.$$

Therefore, the pseudo inverse is:

$$A^\dagger = (A^t A)^{-1} A^t = \begin{pmatrix} -1.2222222 & -1.1111111 & 1.777778 \\ 0.7777778 & 0.8888889 & -1.222222 \end{pmatrix}.$$

Exercise 7.1

Denote $AA^\dagger = E$.

(a) Show that for every $k \in N$, $E^k = E$. Such matrices are called *idempotent* matrices.

(b) Show that $(I - E)E = E(I - E) = 0$.

(c) Show that E is symmetric.

It is clear that if A is a square matrix then the existence of $A^\dagger$ is equivalent to the existence of A^{-1} (and they coincide), so the interest in $A^\dagger$ is mainly in case where $m > n$. Namely, A is a truly tall matrix (with linearly independent columns), in which case $A^\dagger$ is referred to as the *pseudo-inverse* of A. It plays an important role in "solving" the system $Ax = b$ in this case. Indeed, if the system of equations $Ax = b$ does not have a solution, namely, b does not lie in the linear subspace spanned by the columns of A, a natural question is then what is the vector in this space which is the closest to b, namely, using terminology introduced in Chapter 3, what is the projection of b on this linear subspace. The answer to this question is given next.

Theorem

Theorem 7.2. *For the case where $A \in R^{m \times n}$ is a full rank tall matrix,*

$$A^\dagger b = \arg \min_{x \in R^n} \|Ax - b\|.$$

In other words, $AA^\dagger b$ is the projection of b on the linear subspace spanned by the columns of A and its residual equals

$$(I - AA^\dagger)b.$$

Finally,

$$b^t(I - AA^\dagger)b = \min_{x \in R^n} \|Ax - b\|^2.$$

Proof. We next prove only the first and third parts of the theorem as the second part then follows immediately from the first. Before doing that, notice that we claim that $A^\dagger b$ is the unique solution to the optimization problem stated here. Denote $A^\dagger b$ by x^* and note that our goal is to show that for any $x \in R^n$, $x \neq x^*$,

$$\|Ax^* - b\| < \|Ax - b\|.$$

Indeed, for any x,

$$||Ax - b||^2 = ||(Ax - Ax^*) + (Ax^* - b)||^2 = ||Ax - Ax^*||^2 + ||Ax^* - b||^2$$
$$+ 2(Ax - Ax^*)^t(Ax^* - b).$$

We next show that the third term equals zero:

$$(Ax - Ax^*)^t(Ax^* - b) = (x - x^*)^t A^t(Ax^* - b) = (x - x^*)^t(A^t Ax^* - A^t b)$$
$$= (x - x^*)^t(A^t AA^\dagger b - A^t b) = (x - x^*)^t(A^t A(A^t A)^{-1} A^t b - A^t b)$$
$$= (x - x^*)^t(A^t b - A^t b) = 0,$$

as claimed.

In order to show uniqueness, we next argue that unless $x = x^*$, $||Ax - Ax^*||^2 > 0$. Aiming for a contradiction, suppose $||Ax - Ax^*|| = 0$ for some $x \neq x^*$. Then, $Ax = Ax^*$ and $A(x - x^*) = \underline{0}$, violating the assumption that the columns of A are linearly independent.

The final statement of the theorem is based on the fact that

$$||Ax^* - b||^2 = ((I - AA^\dagger)b)^t(I - AA^\dagger)b = b^t(I - AA^\dagger)^t(I - AA^\dagger)b$$
$$= b^t(I - AA^\dagger)b,$$

which can be shown by some algebra.   □

For exercising, you may like to check that the projection and its residual are indeed orthogonal. In fact, we also know that the residual is orthogonal to all columns of A. In the notation we used above,

$$A^t(b - Ax^*) = \underline{0} \in R^m. \tag{7.1}$$

Indeed,

$$A^t(I - AA^\dagger) = A^t - A^t A(A^t A)^{-1} A^t = A^t - A^t = \underline{0} \in R^{m \times m}.$$

Finally, note that Item 3 of Theorem 7.1 is a special case of the above theorem. Indeed, it deals with the case where b lies in the linear subspace spanned by the columns, namely b and its projection coincides. In particular, the residual of the projection of b is the zero vector.

Remark.[b] The optimization problem solved in Theorem 7.2 is equivalent to minimizing

$$\frac{1}{2}||Ax - b||^2 = \frac{1}{2}x^t A^t Ax - x^t A^t b + \frac{1}{2}||b||^2.$$

The gradient of this function equals

$$A^t Ax - A^t b.$$

Equating it to zero, we conclude, as above, that $x^* = (A^t A)^{-1} A^t b$. The Hessien of this matrix is $A^t Ax$ which is a positive matrix (see Section 10.2) as $x^t (A^t A)x = ||Ax||^2 > 0$ for any $x \neq \underline{0}$.

The casting out algorithm (cont.). Note that Theorem 7.2 leads to a procedure for finding if a set of linear equations $Ax = b$, where A is a full-rank matrix, is feasible or no. Of course, this is equivalent to asking whether or not the vector b lies in the linear subspace spanned by the columns of A. Indeed, the answer for this question is yes or no depending whether or not $AA^\dagger b = b$. A possible application for this result is the casting out algorithm, introduced in Section 2.4, for selecting a subset of vectors out of a given set of vectors which forms a basis for the linear subspace spanned by the set of vectors. The idea there was to check whether or not a candidate for entering the basis lies in the linear subspace spanned by those who are already in the basis. Since those that are already in the basis are linearly independent, the matrix A having these vectors as its columns is a full-rank one, we can check whether $AA^\dagger v^i = v^i$ or not, where v^i is the candidate for entering the basis. Indeed, it enters if and only if the answer is no.

An alternative way to derive $A^\dagger$ is via the QR factorization of A:

Theorem

Theorem 7.3. *Let QR be the QR factorization of the full-rank tall (or square) matrix A. Then,*

$$A^\dagger = R^{-1} Q^t. \tag{7.2}$$

---

[b]This remark is for those who are equipped with some calculus background. Otherwise, it can be skipped without loss of continuity.

Proof.

$$A^\dagger = (A^t A)^{-1} A^t = ((QR)^t QR)^{-1} (QR)^t = (R^t Q^t QR)^{-1} R^t Q^t$$
$$= (R^t R)^{-1} R^t Q^t = R^{-1} (R^t)^{-1} R^t Q^t = R^{-1} Q^t,$$

as required. □

Note that this theorem was already derived in (6.3) for the special case where A is an invertible matrix.

Exercise 7.2

Assume $A \in R^{m \times n}$ with $m \geq n$ is a full-rank matrix. Show that $A^\dagger \in R^{n \times m}$ uniquely obeys for $X \in R^{n \times m}$ the following four conditions: $AXA = A$, $XAX = X$, $(AX)^t = AX$ and $(XA)^t = XA$.

Exercise 7.3

Let $Q^{m \times n}$ be an orthonormal matrix with $n \leq m$.
(a) Show that $\sum_{i=1}^n q_i q_i^t = QQ^t$, where q_i is the ith column of Q.
(b) Assume that $A \in R^{m \times n}$, $m \geq n$ and that the columns of A are linearly independent. Show that $AA^\dagger = QQ^t$, where Q is the first term in the QR factorization of A.
(c) Conclude that $\|QQ^t b - b\| = \min_x \|Ax - b\|$.
(d) Show that QQ^t is the projection matrix of the subspace spanned by A's columns, and that $I - QQ^t$ is the residual matrix.

Exercise 7.4

Let

$$A = \begin{pmatrix} -3 & -4 \\ 4 & 6 \\ 1 & 1 \end{pmatrix}, \quad b = \begin{pmatrix} 1 \\ -2 \\ 0 \end{pmatrix}$$

(a) Compute the QR factorization of A.
(b) Compute $A^\dagger$.
(c) Check if there exists a solution to $Ax = b$, and if so, then $x = A^\dagger b$ is the unique solution.

7.2 Simple Linear Regression

We now look at the special case where $n = 2$ and the first column of A is full of ones. This case was already defined and solved completely in Example 2 in Section 2.4. We redo it here as an example for what appears in the previous section. Recall that what we are looking for here is a line $y = ax + b$, which minimizes $\Sigma_{i=1}^n (y_i - ax_i - b)^2$. To align the notation used here with what was previously used, we denote A_{i2} by x_i, $1 \le i \le m$, the decision variables by (b, a), and the right-hand side vector b by y_i, $1 \le i \le n$. Thus,

$$
A = \begin{pmatrix} 1 & x_1 \\ 1 & x_2 \\ \vdots & \vdots \\ 1 & x_m \end{pmatrix} \quad \text{and} \quad b = \begin{pmatrix} y_1 \\ y_2 \\ \vdots \\ y_m \end{pmatrix}.
$$

Then,

$$
A^t A = \begin{pmatrix} 1 & 1 & \cdots & 1 \\ x_1 & x_2 & \cdots & x_m \end{pmatrix} \begin{pmatrix} 1 & x_1 \\ 1 & x_2 \\ \vdots & \vdots \\ 1 & x_m \end{pmatrix} = \begin{pmatrix} m & \sum_{i=1}^m x_i \\ \sum_{i=1}^m x_i & \sum_{i=1}^m x_i^2 \end{pmatrix},
$$

and

$$
(A^t A)^{-1} = \frac{1}{m \sum_{i=1}^m x_i^2 - (\sum_{i=1}^m x_i)^2} \begin{pmatrix} \sum_{i=1}^m x_i^2 & -\sum_{i=1}^m x_i \\ -\sum_{i=1}^m x_i & m \end{pmatrix}
$$

$$
= \frac{1}{m \mathrm{Var}(x)} \begin{pmatrix} \overline{x^2} & -\overline{x} \\ -\overline{x} & 1 \end{pmatrix},
$$

where $\overline{x} = \frac{\sum_{i=1}^m x_i}{m}$ and where $\overline{x^2} = \frac{\sum_{i=1}^m x_i^2}{m}$, coupled with the fact that $\mathrm{Var}(x) = \overline{x^2} - \overline{x}^2$. Then,

$$
A^\dagger = (A^t A)^{-1} A^t = \frac{1}{m \mathrm{Var}(x)} \begin{pmatrix} \overline{x^2} & -\overline{x} \\ -\overline{x} & 1 \end{pmatrix} \begin{pmatrix} 1 & 1 & \cdots & 1 \\ x_1 & x_2 & \cdots & x_m \end{pmatrix}
$$

$$
= \frac{1}{m \mathrm{Var}(x)} \begin{pmatrix} \overline{x^2} - x_1 \overline{x} & \overline{x^2} - x_2 \overline{x} & \cdots & \overline{x^2} - x_m \overline{x} \\ x_1 - \overline{x} & x_2 - \overline{x} & \cdots & x_m - \overline{x} \end{pmatrix}. \tag{7.3}
$$

Finally,

$$
A^\dagger y = \frac{1}{m\operatorname{Var}(x)}
\begin{pmatrix}
\overline{x^2} - x_1\overline{x} & \overline{x^2} - x_2\overline{x} & \cdots & \overline{x^2} - x_m\overline{x} \\
x_1 - \overline{x} & x_2 - \overline{x} & \cdots & x_m - \overline{x}
\end{pmatrix}
\begin{pmatrix}
y_1 \\ y_2 \\ \vdots \\ y_m
\end{pmatrix}
$$

$$
= \frac{1}{m\operatorname{Var}(x)}
\begin{pmatrix}
\overline{x^2}\sum_{i=1}^{m} y_i - \overline{x}\sum_{i=1}^{m} x_i y_i \\
\sum_{i=1}^{m} x_i y_i - \overline{x}\sum_{i=1}^{m} y_i
\end{pmatrix}
= \frac{1}{\operatorname{Var}(x)}
\begin{pmatrix}
\overline{x^2}\,\overline{y} - \overline{x} \times \overline{xy} \\
\operatorname{Cov}(x, y)
\end{pmatrix},
$$

as $\operatorname{Cov}(x, y) = \overline{(x - \overline{x}\underline{1})(y - \overline{y}\underline{1})} = \overline{xy} - \overline{x}\,\overline{y}$. The second entry here, a which is the slope, is easily seen to equal $\operatorname{Corr}(x, y)\frac{\operatorname{SD}(y)}{\operatorname{SD}(x)}$, while after some algebra it can be seen that the first entry, b, which is the intercept, equals $\overline{y} - a\overline{x}$. This is called the *regression line* of Y on X. This line is one of the pillars of statistics and data science. Among others, it allows a quick and dirty function relating two variables. Also, it assists in predicting (albeit with an error) the Y value of a new entry that its X is given.

If we denote by $e \in R^m$ the residual vector, namely $e_i = y_i - ax_i - b$, $1 \leq i \leq m$, we learn from (7.1), that $\Sigma_{i=1}^{m} e_i = 0$ and that $\Sigma_{i=1}^{m} x_i e_i = 0$. Denote $ax_i + b$ by $\hat{y}_i$, $1 \leq i \leq m$, known as the *fitted* values. Then, the last two equations lead immediately to $\Sigma_{i=1}^{m} \hat{y}_i e_i = 0$.

Example. Consider the following data taken from [7, p. 591]:

$$
A = \begin{pmatrix}
1 & 4 \\
1 & 6 \\
1 & 2 \\
1 & 5 \\
1 & 7 \\
1 & 6 \\
1 & 3 \\
1 & 8 \\
1 & 5 \\
1 & 3 \\
1 & 1 \\
1 & 5
\end{pmatrix}
\quad \text{and} \quad
b = \begin{pmatrix}
197 \\
272 \\
100 \\
228 \\
327 \\
279 \\
148 \\
377 \\
238 \\
142 \\
66 \\
239
\end{pmatrix}.
$$

Then,

$$A^t A = \begin{pmatrix} 12 & 55 \\ 55 & 299 \end{pmatrix} \quad \text{and} \quad (A^t A)^{-1} = \frac{1}{459} \begin{pmatrix} 299 & -55 \\ -55 & 12 \end{pmatrix}.$$

Finally, $a = 44.41384$ and $b = 14.186652$. It is straightforward to check that $A^t A = I$ as required.

We next exemplify formula (7.2) for the computation of $A^\dagger$ in this case. Recall from (5.3) that the QR factorization is

$$Q = \frac{1}{\sqrt{m}} \begin{pmatrix} 1 & \frac{x_1 - \bar{x}}{\text{SD}(x)} \\ 1 & \frac{x_2 - \bar{x}}{\text{SD}(x)} \\ \vdots & \vdots \\ 1 & \frac{x_m - \bar{x}}{\text{SD}(x)} \end{pmatrix} \quad \text{and} \quad R = \sqrt{m} \begin{pmatrix} 1 & \bar{x} \\ 0 & \text{SD}(x) \end{pmatrix}.$$

Then,

$$R^{-1} = \frac{1}{\sqrt{m}} \begin{pmatrix} 1 & -\frac{\bar{x}}{\text{SD}(x)} \\ 0 & \frac{1}{\text{SD}(x)} \end{pmatrix},$$

and

$$A^\dagger = R^{-1} Q^t = \frac{1}{m} \begin{pmatrix} 1 & -\frac{\bar{x}}{\text{SD}(x)} \\ 0 & \frac{1}{\text{SD}(x)} \end{pmatrix} \begin{pmatrix} 1 & 1 & \cdots & 1 \\ \frac{x_1 - \bar{x}}{\text{SD}(x)} & \frac{x_2 - \bar{x}}{\text{SD}(x)} & \cdots & \frac{x_m - \bar{x}}{\text{SD}(x)} \end{pmatrix},$$

which is easily seen to coincide with (7.3).

7.3 Multiple Linear Regression

As in the previous section, suppose we sample m individuals but now each possesses values for $p + 1$ variables, where p of which are the explaining variables, denoted by $X^1, \ldots, X^p$ and one is the explained variable, Y. Thus, our data is $X = (X_i^1, X_i^2, \ldots, X_i^p, Y_i)$, $1 \le i \le m$. Each of the variables yields a numerical value across individuals. In many cases, as for example in the previous section, $X_i^1 = 1$, $1 \leq i \leq m$. The matrix X is usually referred to as the *design matrix*.

What we look for here is the set of p values

$$(w_1^*, \ldots, w_p^*) = \arg \min_{w_1, \ldots, w_p} \sum_{i=1}^{m} \left(Y_i - \sum_{j=1}^{p} w_j X_i^j \right)^2 .$$

Assume that $p < m$ as otherwise we usually find a perfect fit and the objective above receives the value of zero. So what we look for is the projection of the vector $Y \in R^m$ on the linear subspace spanned by the vectors $X^1, \ldots, X^p \in R^m$. We assume that X is a full rank matrix, namely its p columns are linearly independent and hence $rank(X) = p$. Finally, the coefficients we are after are

$$w^* = X^\dagger Y,$$

where $X^\dagger = (X^t X)^{-1} X^t$. In particular, w^* is uniquely defined.

The line $\Sigma_{j=1}^p w_j^* X_j$ is known as the *non-constant multiple regression line* of Y on the set of variables $X_1, \ldots, X_p$. When X_1 equals constantly to 1 it is the *multiple regression line* of Y on the set of variables $X_2, \ldots, X_p$. As in the simple regression line, this line serves as an idea on how the explained variable Y is related to the unexplained ones $X_2, \ldots, X_p$, and how the value of the explained variable can be predicted once a new entry with the unexplained variables is given. Additionally, w_j has an interpretation of a derivative. Indeed, it reflects the change of the explained variable due to a unit change in the jth explaining variable, given all other explaining variables are kept unchanged, $1 \leq j \leq p$.

Example. Our goal is to perform multiple linear regression on the following data matrix and observations:

$$X = \begin{pmatrix} 1 & 2 & 1 & 3 \\ 1 & 3 & -2 & 4 \\ 1 & 2 & 1 & 2 \\ 1 & 1 & 2 & -1 \\ 1 & 6 & -1 & 3 \end{pmatrix}, \quad Y = \begin{pmatrix} 1 \\ 2 \\ 3 \\ 4 \\ 5 \end{pmatrix}.$$

We need to compute $X^\dagger Y = (X^t X)^{-1} X^t Y$. Note that by standard computation we get:

$$X^t X = \begin{pmatrix} 5 & 14 & 1 & 11 \\ 14 & 54 & -6 & 39 \\ 1 & -6 & 11 & -8 \\ 11 & 39 & -8 & 39 \end{pmatrix} \implies (X^t X)^{-1}$$

$$= \begin{pmatrix} 2.4995301 & -0.395206767 & -0.7852444 & -0.470864662 \\ -0.3952068 & 0.131109023 & 0.1094925 & 0.002819549 \\ -0.7852444 & 0.109492481 & 0.3566729 & 0.185150376 \\ -0.4708647 & 0.002819549 & 0.1851504 & 0.193609023 \end{pmatrix}.$$

On the other hand,

$$X^t Y = \begin{pmatrix} 15 \\ 48 \\ 3 \\ 28 \end{pmatrix}.$$

Altogether, we get:

$$w^* = X^\dagger Y = \begin{pmatrix} 2.9830827 \\ 0.7725564 \\ -0.2687970 \\ -0.9511278 \end{pmatrix}.$$

Chapter 8

Determinants

8.1 Permutations

Prior to defining the determinant of a square matrix, we need to deal with *permutation*. You may have encountered this concept in a course in combinatorics or probability. The point of departure is the set of n integers $(1, 2, \ldots, n)$, ordered in the natural way. When you permute them, in one way or another, you get a permutation. For example, you can permute $(1, 2, 3)$ to $(2, 3, 1)$, to $(1, 3, 2)$ or even to $(1, 2, 3)$. There are altogether six possible permutations, or $n!$ in the general case. An individual permutation is usually denoted by σ and the set of all permutations of the first n integers is denoted by S_n.

An individual permutation $\sigma \in S_n$ can be looked at as a function which assigned for any integer i, $1 \leq i \leq n$, another integer, say $\sigma(i)$, with $\sigma(i) \in \{1, 2, \ldots, n\}$. Moreover, this is a one-on-one function in the sense that if $i \neq j$, then $\sigma(i) \neq \sigma(j)$. In particular, any j, $1 \leq j \leq n$, comes with a unique i, $1 \leq i \leq n$, where $\sigma(i) = j$. This leads to the inverse permutation, denoted by σ^{-1}. Thus, if $\sigma(i) = j$, then $\sigma^{-1}(j) = i$.

Example. Consider the permutation $\sigma \in S_5$:

$$\sigma = (3, 4, 1, 2, 5). \tag{8.1}$$

Note that σ means that $\sigma(1) = 3$, $\sigma(2) = 4$, $\sigma(3) = 1$, $\sigma(4) = 2$ and $\sigma(5) = 5$. It is easy to see that $\sigma^{-1}(1) = 3$, $\sigma^{-1}(2) = 4$, $\sigma^{-1}(3) = 1$, $\sigma^{-1}(4) = 2$ and $\sigma^{-1}(5) = 5$, or in short

$$\sigma^{-1} = (3, 4, 1, 2, 5).$$

As it turns out in this example $\sigma^{-1} = \sigma$ but this is not the case in general. For example the inverse of $(2, 3, 1)$ is $(3, 1, 2)$.

Note that $\sigma = (\sigma^{-1})^{-1}$, namely the inverse permutation of the inverse permutation is the original permutation. In particular, applying these two permutations consecutively (in any of the two possible orders) is equivalent to applying the identity permutation, namely the permutation $(1, 2, \ldots, n)$. A special case of a permutation σ is a *transposition*. Here for a given $i \neq j$, $\sigma(i) = j$, $\sigma(j) = i$ and $\sigma(k) = k$, $k \neq i, j$. Note that for any transposition σ, $\sigma^{-1} = \sigma$. It is possible to apply one permutation after another. The end result is also a permutation. If $\sigma, \tau \in S_n$, we denote their composition, which is itself a permutation, by $\tau \circ \sigma$, where here σ is applied first and then τ. There is no commutativity in general, however $\sigma^{-1} \circ \sigma = \sigma \circ \sigma^{-1}$ and both equal the identity permutation.

Take two integers i and j and assume without loss of generality that $1 \leq i < j \leq n$. Note that there are as many as $n(n-1)/2$ such pairs. Since a permutation is a one-on-one function, one and only one of the following is possible for any given permutation σ: either $\sigma(i) < \sigma(j)$ or $\sigma(i) > \sigma(j)$.

Definition

Definition 8.1. A permutation $\sigma \in S_n$ is said to be a positive permutation and its sign, denoted by $sgn(\sigma)$ is set to 1, if the number of (i, j) pairs out of the total $n(n-1)/2$ pairs with $1 \leq i < j \leq n$, such that

$$\sigma(i) > \sigma(j) \tag{8.2}$$

is even. Otherwise, it is said to be a negative permutation with $sgn(\sigma) = -1$.

Example (cont.). Consider the permutation σ stated in (8.1). Check all ten pairs i and j with $1 \leq i < j \leq 5$, and find exactly four for which $\sigma(i) > \sigma(j)$, and hence, $sgn(\sigma) = 1$.

Note that the identity permutation, namely the permutation $\sigma \in S_n$ with $\sigma(i) = i$, $1 \leq i \leq n$, is a positive permutation as zero is an even number. It is possible to see that if σ is a transposition then $sgn(\sigma) = -1$.[a]

---

[a]**Proof.** Suppose $i < j$ and consider the transposition $(1, 2, \ldots, i-1, j, i+1, \ldots, j-1, i, j+1, \ldots, n)$. All $j - i - 1$ numbers which are strictly between i and j, when compared to both i and j, contribute two integers where the inequality is reversed. Adding to it the one inequality between i and j, and we get $2(j - i - 1) + 1$, which is an odd number.

We also claim, without a proof, that $sgn(\tau \circ \sigma) = sgn(\tau)sgn(\sigma)$. Hence, for any permutation $\sigma \in S_n$, $sgn(\sigma^{-1}) = sgn(\sigma)$.

The exercise below and its solution are given in [8].

For the following permutations:

$$\sigma_1 = \begin{pmatrix} 1 & 2 & 3 & 4 & 5 \\ 2 & 4 & 5 & 1 & 3 \end{pmatrix}, \quad \sigma_2 = \begin{pmatrix} 1 & 2 & 3 & 4 & 5 \\ 4 & 1 & 3 & 5 & 2 \end{pmatrix}.$$

(a) Compute σ^{-1} and $\sigma_1 \circ \sigma_2$.

(b) For every permutation, determine its sign (including permutations computed in (a)).

(c) Show that in general $sgn(\sigma) = sgn(\sigma^{-1})$.

8.2 The Determinant

Throughout this chapter we consider only square matrices so this fact is not going to be repeated. Likewise, when we refer to the product AB we assume that is it well-defined.

Definition

Definition 8.2. Let $A \in R^{n \times n}$. The determinant of A, denoted by $det(A)$, is defined by

$$det(A) = \sum_{\sigma \in S_n} \prod_{i=1}^{n} sgn(\sigma)A_{i\sigma(i)}. \tag{8.3}$$

Examples.

- For a scalar a, $det(a) = a$.
- For

$$A = \begin{pmatrix} a & b \\ c & d \end{pmatrix},$$

$det(A) = ad - bc$. You may recall this term which was used when the inverse of a 2×2 matrix was defined in Chapter 6.

- Let $A \in R^{3 \times 3}$. Since $3! = 6$, there are six permutations to be considered. The three permutations $(1, 2, 3)$, $(2, 3, 1)$ and $(3, 1, 2)$ are positive, while the three permutations $(3, 2, 1)$, $(2, 1, 3)$ and $(1, 3, 2)$ are negative. Hence,

$$det(A) = A_{11}A_{22}A_{33} + A_{12}A_{23}A_{31} + A_{13}A_{21}A_{32}$$
$$- A_{13}A_{22}A_{31} - A_{12}A_{21}A_{33} - A_{11}A_{23}A_{32}.$$

- $det(A^t) = det(A)$.

  Proof. When we run through all the permutations and the corresponding products when $det(A^t)$ is computed, we in fact get the same set of products, but the factor that multiplied $sgn(\sigma)$ when $det(A)$ was computed, now multiplies $sgn(\sigma^{-1})$. Since $sgn(\sigma^{-1}) = sgn(\sigma)$ the proof is completed. □

- If A is an upper or lower triangular matrix, then its determinant equals the product of the entries along the diagonal. This is the case since with exception of the identity permutation, all corresponding product comes with at least one zero. Finally, if additionally, one of the diagonal entries equals zero (and hence the matrix is not invertible), then also the product which corresponds to the identity permutation equals zero. Also, $det(I) = 1$.

Exercise 8.2

Compute the determinants of the following matrices:

$$A = \begin{pmatrix} 2 & -3 \\ 4 & 7 \end{pmatrix}, \quad B = \begin{pmatrix} 1 & -2 & 3 \\ 2 & 4 & -1 \\ 1 & 5 & -2 \end{pmatrix}, \quad C = \begin{pmatrix} 1/2 & -1 & -1/3 \\ 3/4 & 1/2 & -1 \\ 1 & -4 & 1 \end{pmatrix}.$$

Exercise 8.3

Claim: for $A \in R^{n \times n}$, we have $det(A) = det(A^t)$. Prove that claim for matrices of dimension 2×2.

8.3 Determinants and Row Operations

In Section 5.6, we have defined three row operations. We next describe what is the effect of each one of them, when applied, to the determinant of a matrix.

- If a row (or a column) is multiplied by a constant c, then the matrix determinant is multiplied by c too.[b]
- If two rows (or columns) are swapped, then the determinant keeps its absolute value but its sign is reversed. This implies that:

Theorem

Theorem 8.1. *If a matrix has two identical rows (or columns), its determinant equals zero.*

- If c times of one row is subtracted from another, the determinant is kept unaltered.

Exercise 8.4

Prove Theorem 8.1 using the three row/column operations and their effect on the determinant.

Exercise 8.5

(a) Let $A \in R^{n \times n}$ and let A' be the matrix obtained by multiplying the ith row of matrix A by a scalar λ. Prove that $det(A') = \lambda det(A)$.
(b) Let $A \in R^{n \times n}$ and let A' be the matrix obtained by swapping two rows of matrix A. Prove that $det(A') = -det(A)$.

Exercise 8.6

Let $A, B, C \in R^{n \times n}$. Suppose that A is identical to B, except the i^*th column, and suppose that C is identical to A, but its i^*th column is the sum $A._{i^*} + B._{i^*}$. Find an expression for $det(C)$ in term of $det(A)$ and $det(B)$.

Recall that applying a row operation on a matrix is equivalent to multiplying it from the left by the relevant matrix. If we denoted these matrices by E_i, $1 \leq i \leq n$ then, keeping the order of the row operations stated above, we get that $det(E_1) = c$, $det(E_2) = -1$ and $det(E_3) = 1$.

---

[b]Note that this fact also holds for the constant zero, although zero was excluded when this row operation was defined.

Theorem

Theorem 8.2. *For any row operation matrix E, $det(EA) = det(E)det(A)$.*

Proof. We need to consider three row operations. Keeping the order in which they were introduced above, the result is trivial for the first among them. As for the second row operation, observe first that among the $n!$ products involved in computing $det(E_2)$, all but one are zero. The remaining one gets an absolute value of one. We need to argue that the sign of the corresponding permutation is -1. Indeed, suppose rows i and j are swapped. Then clearly the permutation we are after is a transposition between i and j. As we showed above the sign of all transpositions is -1. Finally, for the third operation. Suppose row j is deducted c times from row i. Thus, wherever A_{ik} appears in any product leading to $det(A)$, we now have $(A_{ik} - cA_{jk})$. Breaking all such products into a sum of two products, one with A_{ik} and another with $-cA_{jk}$ we get that the former equals $det(A)$ while the latter is in fact c times a determinant of a matrix with two identical rows, as the original jth row now appears twice. This determinant equals zero. $\square$

Suppose a matrix B can be reached from A by applying a number of row operations, called *row equivalence*, namely

$$B = E_k E_{k-1} \cdots E_2 E_1 A,$$

for some series of row operations matrices, $E_1, E_2, \ldots, E_k$, $B = E_k E_{k-1} \cdots E_2 E_1 A$. Then, from the last theorem we learn that

$$det(B) = \prod_{i=1}^{k} det(E_i)det(A).$$

Since I can be reached from A if and only if A is invertible, and since $det(I) = 1$, we conclude:

Theorem

Theorem 8.3. A^{-1} *exists if and only if $det(A) \neq 0$. In which case $det(A^{-1}) \neq 0$ too.*

Theorem

Theorem 8.4. $det(AB) = det(A)det(B)$

Proof. The case where A is not invertible was already proved in Section 6.5 that the same is the case where AB. Hence in this case $det(AB)$ and $det(A) = 0$, making $det(AB) = det(A)det(B)$ immediate. So assume now that A is invertible. Hence, $I = E_k \cdots E_1 A$ for some series of row operation matrices $E_1, \ldots, E_k$. Then, $A = E_1^{-1} \cdots E_k^{-1}$. But the inverse matrices in this product are themselves row operations matrices. Hence, $A = F_1 \cdots F_k$ for some row operation matrices $F_1, \ldots, F_k$. Hence, by Theorem 8.2, $det(A) = \prod_{i=1}^{k} det(F_i)$. Then, $AB = \prod_{i=1}^{k} F_i B$. Applying Theorem 8.2 again we get that $det(AB) = \prod_{i=1}^{k} det(F_i) det(B)$. The fact that $det(A) = \prod_{i=1}^{k} det(F_i)$ concludes our proof. $\square$

The following three conclusions are now immediate:

Theorem

Theorem 8.5.

- *if A is invertible, then $det(A^{-1}) = 1/det(A)$.*
- *if A and B are similar, then $det(B) = det(A)$.*
- *for a unitary matrix U, $det(U) = 1$.*

8.4 Minor Matrices and the Determinant

In (8.3), the definition for $det(A)$ for $A^{n \times n}$ is given which is constructive in the sense that it states a formula for computing it. We next deal with an alternative computation scheme. It is recursive in nature in the sense that a determinant of an $n \times n$ matrix is stated in terms of determinants of (related) $(n-1) \times (n-1)$ matrices. They, in turn, can be computed using determinants of (related) $(n-2) \times (n-2)$ matrices, etc., going all the way to determinants of scalars which coincide with their values. The procedure thus commences with determinants of 1×1 matrices, which are scalars. As it turns out, these scalars are the n^2 entries of the matrix. The main result is stated next and we see later that it also has a theoretical merit. But first we need the following definition.

Definition

Definition 8.3. The M_{ij} minor of a matrix $A \in R^{n \times n}$ is an $(n-1) \times (n-1)$ matrix which is derived from A by removing its ith row and its jth column, $1 \leq i, j \leq n$.

Theorem

Theorem 8.6. *Let* $A \in R^{n \times n}$. *Then,*

$$det(A) = \sum_{j=1}^{n} (-1)^{i+j} A_{ij} det(M_{ij}), \quad 1 \leq i \leq n. \tag{8.4}$$

Proof. Suppose $i = 1$ and consider the entry A_{11}. Factor it out from all products which contain it when $det(A)$ is computed. You will see that it multiplies M_{11}. Next factor out $A_{12}, A_{13}, \ldots, A_{1n}$. First, note that all products in the summation defining $det(A)$ appear exactly once. What, for example, multiplies A_{12}? Had it been in position $(1,1)$ in the case where the first two columns have been swapped the answer clearly would have been M_{12}. But this swapping implies that any permutation was composite with a transposition and hence its sign has been changed from 1 to -1 or from -1 to 1. Thus, in fact, A_{12} is multiplied by $-M_{12}$. Next consider M_{13}. A similar argument holds here but now two transpositions are involved (column 3 and column 2 were firstly swapped and (the new) columns 2 and 1 were swapped next), keeping the sign unaltered. A similar argument holds for all other entries in this row. Our proof is now complete for the case where $i = 1$. For any other value for i, note that in order to place A_{ij} at the $(1,1)$ position, the total number of rows and column swaps needed is $i + j - 2$, and hence the term $(-1)^{i+j} = (-1)^{i+j-2}$ appears is the product. $\qquad \square$

The method for computing the determinant as it appears in Theorem 8.6 and formula (8.4) is known as developing the determinant along (or around) a particular row.

Example. Developing the determinant along the second row leads to:

$$det \begin{pmatrix} 3 & 1 & -2 \\ 4 & 2 & 5 \\ 1 & -3 & 0 \end{pmatrix} = -4det \begin{pmatrix} 1 & -2 \\ -3 & 0 \end{pmatrix} + 2det \begin{pmatrix} 3 & -2 \\ 1 & 0 \end{pmatrix} - 5det \begin{pmatrix} 3 & 1 \\ 1 & -3 \end{pmatrix}$$

$$= -4 \times (-6) + 2 \times 2 - 5 \times (-9 - 1) = 78.$$

Of course, if we elect to develop the determinant along some other row (or column), the same will be the end result.

Since $det(A^t) = det(A)$, we can replace the role of a row by that of a column in (8.4). Specifically,

$$det(A) = \sum_{i=1}^{n} (-1)^{i+j} A_{ij} det(M_{ij}), \quad 1 \le j \le n.$$

Suppose one applies formula (8.4) but with a wrong row of A. Thus, A_{ij} is replaced with A_{kj} for some $k \ne i$ (but the minors M_{ij} stay untouched). This is equivalent to computing the determinant of the matrix which is the same as A but its ith row is replaced by the kth. Since the kth row remains as is, we in fact compute the determinant of a matrix with two identical rows. This determinant does not call for computation: We know that it equals zero. In summary,

$$det(A) = \sum_{j=1}^{n} (-1)^{i+k} A_{kj} det(M_{ij}) = 0, \quad 1 \le k \ne i \le n. \qquad (8.5)$$

Of course, a similar result holds for columns:

$$det(A) = \sum_{i=1}^{n} (-1)^{i+k} A_{ik} det(M_{ij}) = 0, \quad 1 \le k \ne j \le n.$$

Exercise 8.7

Compute the determinant of the following matrices:

$$A = \begin{pmatrix} 2 & 0 & 1 \\ 2 & 0 & 5 \\ 3 & 7 & 2 \end{pmatrix}, \quad B = \begin{pmatrix} 5 & 7 & 0 & 1 \\ 3 & 1 & 0 & 3 \\ 1 & 1 & 1 & 2 \\ 3 & 0 & 0 & 4 \end{pmatrix}, \quad C = \begin{pmatrix} 0 & 1 & 0 & 0 \\ 1 & 0 & 1 & 1 \\ 1 & 1 & -1 & 0 \\ 1 & 1 & 0 & -1 \end{pmatrix}$$

Example (The Vandermonde determinant). Let x_i, $1 \leq i \leq n$, be a set of n real numbers. Consider the following $n \times n$ matrix:

$$A = \begin{pmatrix} 1 & x_1 & x_1^2 & \cdots & x_1^{n-1} \\ 1 & x_2 & x_2^2 & \cdots & x_2^{n-1} \\ \vdots & \vdots & \vdots & & \vdots \\ 1 & x_n & x_n^2 & \cdots & x_n^{n-1} \end{pmatrix}. \tag{8.6}$$

We claim that the determinant of this matrix equals

$$det(A) = \prod_{i>k}(x_i - x_k). \tag{8.7}$$

For the case where two (or more) of the x_i's in this set are equal, this product equals zero. Indeed, in this case two rows in the matrix A are identical, a case where we already know that the determinant equals zero. For a proof for the general case we will use an induction argument on n. For the case where $n = 1$ we get that the matrix is the scalar one, making this the determinant. Also, product (8.7) is now empty, which by definition, sets it to 1. For good measure look also at the case of $n = 2$, where it is easy to see that the product now equals $x_2 - x_1$, as required. Consider now the general case. We can make the first row to be $(1, 0, \ldots, 0)$ by a series of column operations. Specifically, this can be done by multiplying the $(n-1)$th column by x_1 and subtracting it from the nth column. Now do that but with the $(n-2)$th and the $(n-1)$th columns, etc. This leads to the matrix

$$\begin{pmatrix} 1 & 0 & 0 & \cdots & 0 \\ 1 & x_2 - x_1 & x_2^2 - x_1 x_1 & \cdots & x_2^{n-1} - x_1 x_2^{n-2} \\ \vdots & \vdots & \vdots & \vdots & \\ 1 & x_n - x_1 & n_1^2 - x_1 x_n & \cdots & x_n^{n-1} - x_1 x_n^{n-2} \end{pmatrix}.$$

The determinant, so far, has not changed. Next we make the first column the e_1 vector. This can be easily done by subtracting the current first row from all other rows. We then get

$$\begin{pmatrix} 1 & 0 & 0 & \cdots & 0 \\ 0 & x_2 - x_1 & x_2^2 - x_1 x_1 & \cdots & x_2^{n-1} - x_1 x_2^{n-2} \\ \vdots & \vdots & \vdots & \vdots & \\ 0 & x_n - x_1 & x_n^2 - x_1 x_n & \cdots & x_n^{n-1} - x_1 x_n^{n-2} \end{pmatrix}.$$

Again, no change in the determinant. Next, if we take the common factor $x_2 - x_1$ out of the second row, $x_3 - x_1$ from the third row, etc., we then get

$$det(A) = \prod_{i=2}^{n}(x_i - x_1)det \begin{pmatrix} 1 & 0 & 0 & \cdots & 0 \\ 0 & 1 & x_2 & \cdots & x_2^{n-2} \\ \vdots & \vdots & \vdots & \vdots & \vdots \\ 0 & 1 & x_n & \cdots & x_n^{n-2} \end{pmatrix}.$$

Consider the determinant of the matrix above, while ignoring the multiplicative constant. By developing it along the first row, we get that it equals one times the determinant of a Vandermonde matrix but with $n-1$ points. Invoking the induction hypothesis, its determinant equals $\Pi_{i>k\geq 2}(x_i - x_k)$. Multiplying this by $\Pi_{i=2}^{n}(x_i - x_1)$ concludes the proof.

Exercise 8.8

Show that

$$det \begin{pmatrix} 1 & 1 & 1 & \cdots & 1 \\ 1 & 2-x & 1 & \cdots & 1 \\ 1 & 1 & 3-x & \cdots & 1 \\ \vdots & \vdots & \vdots & \ddots & \vdots \\ 1 & 1 & 1 & \cdots & n+1-x \end{pmatrix} = \prod_{i=1}^{n}(i-x).$$

Hint: Use row/column operations to obtain a triangular matrix.

Exercise 8.9

Show that

$$det \begin{pmatrix} sin^2\alpha & 1 & cos^2\alpha \\ sin^2\beta & 1 & cos^2\beta \\ sin^2\gamma & 1 & cos^2\gamma \end{pmatrix} = 0$$

using the fact that $sin^2\alpha + cos^2\alpha = 1$.

8.5 The Adjoint Matrix

This brings us to defining a matrix associated with matrix A, which is called its *adjoint* and is denoted by $adj(A)$:

Definition

Definition 8.4. For a matrix $A \in R^{n \times n}$, denote its adjoint matrix $adj(A) \in R^{n \times n}$ by

$$[adj(A)]_{ij} = (-1)^{i+j} det(M_{ji}), \quad 1 \le i, j \le n.$$

Note that the order of the indices in M_{ji} above is not an error.

Theorem

Theorem 8.7. *For any matrix A,*

$$A adj(A) = det(A)I = adj(A)A.$$

In particular, if A is invertible, then

$$A^{-1} = \frac{1}{det(A)} adj(A). \tag{8.8}$$

Proof. The (i,j)th entry in the product $A adj(A)$ equals

$$\sum_{k=1}^{n} A_{ik}[adj(A)]_{kj} = \sum_{k=1}^{n} A_{ik}(-1)^{k+j} det(M_{jk}), \quad 1 \le i, j \le n.$$

In the case where $i = j$, we get $det(A)$ as we are in effect developing it along the ith row (see (8.4)). Otherwise, by (8.5), we get zero. This concludes the proof for the first equality. The second equality can be shown in a similar way. $\qquad \square$

Example. The adjoint of

$$A = \begin{pmatrix} 2 & 1 & 3 \\ 0 & 1 & 2 \\ 1 & 0 & -2 \end{pmatrix} \quad \text{is} \quad \begin{pmatrix} -2 & 2 & -1 \\ 2 & -7 & -4 \\ -1 & 1 & 2 \end{pmatrix}.$$

Then,

$$Aadj(A) = \begin{pmatrix} -5 & 0 & 0 \\ 0 & -5 & 0 \\ 0 & 0 & -5 \end{pmatrix}.$$

In particular, $det(A) = -5$.

Exercise 8.10

Let $A \in R^{n \times n}$.
(a) Prove that for any scalar c, $det(cA) = c^n det(A)$.
(b) Prove that $det(adj(A)) = det(A)^{n-1}$.

Exercise 8.11

Let:

$$A = \begin{pmatrix} 2 & 3 & 4 \\ 5 & 6 & 7 \\ 8 & 9 & 1 \end{pmatrix}, \quad B = \begin{pmatrix} 2 & 3 & 4 \\ 5 & 4 & 3 \\ 1 & 2 & 1 \end{pmatrix}.$$

(a) Compute the adjoint matrix of each.
(b) Compute the inverse matrix of each.

8.6 Cramer's Method for Solving Linear Equations

We deal with one more (and last) technique for solving a linear system of equations $Ax = b$ where A is invertible. It is known as Cramer's method.

Theorem

Theorem 8.8. *Consider the system of linear equations $Ax = b$ where $A \in R^{n \times n}$ is an invertible matrix. Let Δ_i be the determinant of the matrix which is the same as A except that its ith column is replaced by the vector b, $1 \le i \le n$. Then, the unique solution for this system is*

$$x_i = \frac{\Delta_i}{det(A)}, \quad 1 \le i \le n. \tag{8.9}$$

Proof. We already know from (8.8) that

$$x = A^{-1}b = \frac{1}{det(A)}adj(A)b.$$

Looking at the ith entry here, we get that

$$x_i = \frac{\sum_{j=1}^{n} b_j [adj(A)]_{ij}}{det(A)} = \frac{\sum_{j=1}^{n} b_j (-1)^{i+j} det(M_{ji})}{det(A)}, \quad 1 \le i \le n.$$

Note that had there been A_{ji} instead of b_j at the last summation, we would get $det(A)$ in the numerator. Thus, what we have there is in fact the determinant of the matrix which is the same as A but its ith column has been replaced with the vector b. $\qquad\square$

Exercise 8.12

Use Cramer's method to solve the following system of linear equations:

$$x_1 + 4x_2 + 2x_3 = 8,$$

$$2x_1 - 3x_2 + x_3 = 12,$$

$$6x_1 + x_2 - 8x_3 = -29.$$

8.6.1 *Fitting a polynomial*

Let (x_i, y_i), $1 \le i \le n$, be n points in the two-dimensional plane. Assume none of the x values are being repeated. It is well-known that it is possible to draw one and only $(n-1)$-degree polynomial which passes through all these points. I'll spoil the show by telling you that this polynomial equals

$$p(x) = \sum_{i=1}^{n} y_i \frac{\prod_{k \ne i}(x - x_k)}{\prod_{k \ne i}(x_i - x_k)}.$$

This is easily checked by inserting x_i in $p(x)$ and noticing that $p(x_i) = y_i$, $1 \le i \le n$. The common denominator of the terms in the summation defining $p(x)$ is $\Pi_{1 \le i < k \le n}(x_i - x_k)$, which is the Vandermonde determinant (see (8.7)). Should this surprise us? The answer is no. Specifically, by definition $p(x) = \sum_{i=0}^{n-1} a_i x^i$ for some coefficients a_i, $0 \le i \le n-1$. Then the vector $\{a_i\}_{i=0}^{n-1}$ solves the system of linear equations $Ax = b$ where A is stated in (8.6) and b comes with the y values, namely $b_i = y_i$, $1 \le i \le n$. If we solve this system using the Cramer method, we should get $det(A)$, namely the Vandemonde determinant, in the denominator of (8.9). As for

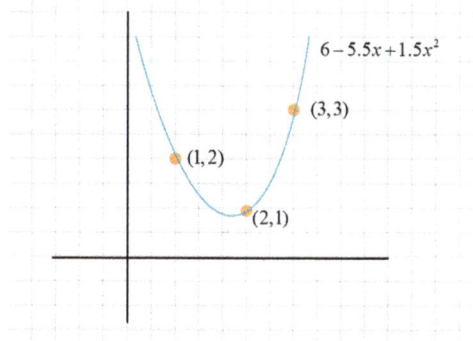

Figure 8.1. Three points of a quadratic function.

the numerator, we do what is prescribed by Cramer's and find the determinant of matrix A where its jth column is replaced with the y vector. This is illustrated in the example below.

Example. Consider the following three points in the two-dimensional plane: $(1, 2)$, $(2, 1)$ and $(3, 3)$. Suppose one looks for a polynomial of degree two, namely a quadratic function, which crosses these three points (see Fig. 8.1). Denote this polynomial by $a_0 + a_1 x + a_2 x^2$. Thus, we look for a_0, a_1 and a_2 which solve

$$\begin{pmatrix} 1 & 1 & 1 \\ 1 & 2 & 4 \\ 1 & 3 & 9 \end{pmatrix} \begin{pmatrix} a_0 \\ a_1 \\ a_2 \end{pmatrix} = \begin{pmatrix} 2 \\ 1 \\ 3 \end{pmatrix}.$$

The determinant of the constraint (Vandermonde) matrix equals $(1 - 2)(1 - 3)(2 - 3) = 2$. Hence,

$$a_0 = \frac{det \begin{pmatrix} 2 & 1 & 1 \\ 1 & 2 & 4 \\ 3 & 3 & 9 \end{pmatrix}}{2} = \frac{12}{2}, \quad a_1 = \frac{det \begin{pmatrix} 1 & 2 & 1 \\ 1 & 1 & 4 \\ 1 & 3 & 9 \end{pmatrix}}{2} = -\frac{11}{2} \quad \text{and}$$

$$a_2 = \frac{det \begin{pmatrix} 1 & 1 & 2 \\ 1 & 2 & 1 \\ 1 & 3 & 3 \end{pmatrix}}{2} = \frac{3}{2}.$$

Computing the above determinants, we conclude that the quadratic polynomial is $6 - 5.5x + 1.5x^2$.

Chapter 9

Eigensystems and Diagonalizability

9.1 The Characteristic Polynomial

> **Definition**
>
> **Definition 9.1.** Assume $A \in R^{n \times n}$. The n degree polynomial in $t \in R$, denoted by $P_A(t)$ and defined by $det(tI - A)$, is called the characteristic polynomial of A.

Example 1. Let

$$A = \begin{pmatrix} 1 & 3 & 0 \\ -2 & 2 & -1 \\ 4 & 0 & -2 \end{pmatrix}.$$

Then,

$$P_A(t) = det(tI - A) = det \begin{pmatrix} t-1 & -3 & 0 \\ 2 & t-2 & 1 \\ -4 & 0 & t+2 \end{pmatrix} = t^3 - t^2 + 2t + 28.$$

Note that $P_A(0) = (-1)^n det(A)$.

9.2   Right and Left Eigensystems

Definition

Definition 9.2. A scalar λ is said to be an eigenvalue of A if it is a root of its characteristic polynomial, namely if $P_A(\lambda) = 0$. Equivalently, if there exists a non-zero vector $v \in R^n$ such that $Av = \lambda v$, then v right eigenvector of A associated with λ, or, equivalently, if there exists a non-zero vector $u \in R^n$, such that $A^t u = \lambda u$, then u is called a left eigenvector of A associated with λ.[a]

---

[a]If one allows scalars and vectors to come with complex numbers, the set of eigenvalues and eignevectors may increase. Yet, we do not consider this option throughout this text.

Example 2. For the matrix $A \in R^{2\times 2}$

$$A = \begin{pmatrix} a & b \\ c & d \end{pmatrix},$$

$$P_A(t) = x^2 - trace(A)t + det(A),$$

where $trace(A)$ is the sum of the entries along the diagonal of A. In the case where $trace^2(A) < 4det(A)$ no eigenvalues exist. Otherwise, the two eigenvalues exist and their values are $(-trace(A) \pm \sqrt{trace^2(A) - 4det(A)})/2$. Note that $trace(A)$ is their sum and $det(A)$ is their product. Of course, the two coincide in the case where $trace^2(A) = 4det(A)$.

It is easy to see that for any given eigenvalue, the corresponding set of right eigenvectors form a linear subspace of R^n. An eigenvalue and its associated eigenvectors are referred to as *eigensystem*. The same can be said about the set of left eigenvectors.

Zero is certainly a candidate for being an eigenvalue, and indeed it is if and only if A is singular, namely it comes with a nonzero vector $x \in R^n$ with $Ax = \underline{0}$. We do not claim that eignevalues exist at all, although we herein state the spoiler that this is the case for symmetric matrices, which is the subject of the next section.

Remark. Computing an eigenpair of an eigenvalue λ and its (right) eigenvector v is not an easy task, even if one knows in advance that such an eigenpair exists. Yet, if one knows λ, then finding v is equivalent for solving a system of linear equations $(A - \lambda I)x = \underline{0}$. If on the other hand v is given, then, as easily checked,

$$\lambda = \frac{v^t A v}{v^t v}.$$

Indeed,

$$\frac{v^t A v}{v^t v} = \frac{v^t \lambda v}{v^t v} = \lambda \frac{v^t v}{v^t v} = \lambda.$$

Another possibility is that more than one eigenvalue exists. In this case, the following result holds.

> **Theorem**
>
> **Theorem 9.1.** *Let λ and μ be two eigenvalues of A. Let v be a right eigenvector associated with λ and u be a left eigenvector associated with μ. If $\mu \neq \lambda$, then v and u are orthogonal.*

Proof. It is easy to see that $u^t A v$ equals to both $\lambda u^t v$ and $\mu u^t v$. This is possible if $\mu = \lambda$ or if u and v are orthogonal (or both). $\qquad \square$

Example 3.[a] Let

$$A = \begin{pmatrix} 1 & 2 \\ 3 & 2 \end{pmatrix} \quad \text{and} \quad P_A(x) = det \begin{pmatrix} x-1 & -2 \\ -3 & x-2 \end{pmatrix}.$$

Thus, $P_A(t) = t^2 - 3t - 4$ which comes with two eigenvalues: $\lambda_1 = 4$ and $\lambda_2 = -1$. In order to find a right eigenvector associated with 4 we need to consider the two linear equations:

$$x_1 + 2x_2 = 4x_1 \quad \text{and} \quad 3x_1 + 2x_2 = 4x_2.$$

Yet, whatever solves the first equation, solves also the second, as expected. A possible solution is $(2, 3)$. In fact, the set of eigenvectors are all scalar multiplications of this vector. In order to find a right eigenvector associated with -1 we need to consider the two linear equations:

$$x_1 + 2x_2 = -x_1 \quad \text{and} \quad 3x_1 + 2x_2 = -x_2.$$

Here all the scalar multiplications of $(1, -1)$ meet these conditions. As for left eigenvectors we look for a solution to two systems of linear equations.

---

[a]This example appears in [8, p. 309].

First,

$$y_1 + 3y_2 = 4y_1 \quad \text{and} \quad 2y_1 + 2y_2 = 4y_2,$$

which is solved by all scalar multiplication of $(1, 1)$. Second,

$$y_1 + 3y_2 = -y_1 \quad \text{and} \quad 2y_1 + 2y_2 = -y_2,$$

which is solved by all scalar multiplication of $(3, -2)$. Note that $(2, 3)$ and $(3, -2)$ are orthogonal. The same is true regarding the pair $(1, -1)$ and $(1, 1)$.

Theorem

Theorem 9.2. *Let λ be an eigenvalue of $A \in R^{n \times n}$ with u and v being, respectively, its right and left eigenvectors. Assume that $u^t v = 1$. Let $\mu \neq \lambda$ be another eigenvector of A. Then, for the matrix $A - \lambda v u^t$:*

1. *0 is an eigenvalue of it with v and u being, respectively, its right and left eigenvectors*
2. *μ is its eigenvalue with the same set of eigenvectors as for A.*

Proof. Left as an exercise. □

Exercise 9.1

(a) Let A, B be two square matrices. Show that both matrices AB and BA have the same eigenvalues.

(b) Show that the matrix $A \in \mathbf{R}^{n \times n}$ has the same characteristic polynomial as A^t.

(c) Show that the set of vectors $v \in R^n$ which satisfy $Av = \lambda v$ for a given $\lambda \in R$, form a linear sub-space of R^n.

(d) Let[b] $A = \begin{pmatrix} 3 & -4 \\ 2 & -6 \end{pmatrix}$. Find all eigenvalues and corresponding eigenvectors of A.

---

[b]This example appears in [8, p. 309].

Exercise 9.2

Let:

$$A = \begin{pmatrix} 1 & 4 & 3 \\ 0 & 3 & 1 \\ 0 & 2 & -1 \end{pmatrix}.$$

(a) What is the characteristic polynomial of A?
(b) Prove that 1 is an eigenvalue of A.
(c) Are there any more eigenvalues? If so, compute them.

9.3 Algebraic and Geometric Multiplicities of Eigenvalues

It is possible to see that λ is an eigenvalue of A if and only if $P_A(t) = (t - \lambda)Q(t)$ for some $n - 1$ degree polynomial $Q(t)$. This does not rule out that λ is also a root of $Q(t)$. In this case $P_A(t) = (t - \lambda)^2 T(t)$ for some $(n - 2)$-degree polynomial $T(t)$. This brings us to the following definition.

Definition

Definition 9.3. The eigenvalue λ of a square matrix $A \in R^{n \times n}$ is said to have an algebraic multiplicity of $m \geq 1$ if $P_A(t) = (t - \lambda)^m Q(t)$ for some polynomial $Q(t)$ of degree $n - m$ with $Q(\lambda) \neq 0$. An eigenvalue is called simple if its algebraic multiplicity equals one.

Clearly, the algebraic multiplicity of an eigenvalue is an integer between 1 and n. Moreover, the sum of the algebraic multiplicities across the eigenvalues needs to be smaller than or equal to n. In particular, if $\lambda_1, \ldots, \lambda_r$ are the eigenvalues whose algebraic multiplicities are $k_1, \ldots, k_r$, respectively, then

$$P_A(t) = \prod_{i=1}^{r} (t - \lambda_i)^{k_i} Q(t),$$

for some polynomial $Q(x)$ of degree $n - \Sigma_{i=1}^{k} r_i$ with $Q(\lambda_i) \neq 0$, $1 \leq i \leq r$. For the special case where $\sum_{i=1}^{r} k_i = n$, one gets the following.

> **Theorem**
>
> **Theorem 9.3.** *Let* $\lambda_1, \ldots, \lambda_r$ *be the eigenvalues of A whose algebraic multiplicities are* $k_1, \ldots, k_r$. *If* $\sum_{i=1}^{r} k_i = n$, *then*
>
> $$P_A(t) = \prod_{i=1}^{r} (t - \lambda_i)^{k_i}.$$
>
> *Moreover, writing* $P_A(t)$ *as* $\Sigma_{j=0}^{n} a_j t^j$, *then* $a_0 = (-1)^n \prod_{i=1}^{r} \lambda_i^{k_i}$, $a_{n-1} = -\Sigma_{i=1}^{r} k_i \lambda_i = \Sigma_{i=1}^{n} A_{ii}$ *(known as the trace of A) and* $a_n = 1$. *Finally,* $det(A) = (-1)^n P_A(0) = (-1)^n a_0 = \prod_{i=1}^{r} \lambda_i^{k_i}$.

Note the condition stated in the above theorem. By no means we say that it always holds or even pretend that it does not hold only in some pathological cases. Yet, as we show in the next chapter, it holds for symmetric matrices.

Next we define another concept related to eigenvalues.

> **Definition**
>
> **Definition 9.4.** The eigenvalue λ of the square matrix $A \in R^{n \times n}$ is said to have a geometric multiplicity of k for some $1 \leq k \leq n$, if $k = dim(null(\lambda I - A))$. Put differently (see Theorem 4.3), if the dimension of the subspace of right (and hence of left) eigenvectors associated with λ equals k.

We next state a theorem whose proof is deferred to a later section.

> **Theorem**
>
> **Theorem 9.4.** *The algebraic multiplicity of an eigenvalue is larger than or equal to its geometric multiplicity. In particular, the geometric multiplicity of a simple eigenvalue is one.*

This theorem is now exemplified.

Example 4. Let

$$A = \begin{pmatrix} 3 & -1 & 1 \\ 7 & -5 & 1 \\ 6 & -6 & 2 \end{pmatrix}.$$

Some algebra leads to $P_A(x) = x^3 - 12x + 16 = (x-2)^2(x-4)$. Thus, there exist two eigenvalues 2 and 4. Their algebraic multiplicities are 2 and 1, respectively. There is no question regarding the geometric multiplicity of the eigenvalue 4: it equals 1. In order to find the geometric multiplicity of the eigenvalue 2 we need to solve the homogeneous system of linear equations:

$$3x_1 - x_2 + x_3 = 2x_1,$$

$$7x_1 - 5x_2 + x_3 = 2x_2.$$

Note that we ignore the third equation $6x_1 - 6x_2 + 2x_3 = 2x_3$ as it is automatically solved by any solutions to the previous two equations. Since we have two linearly independent equations with three unknowns, the dimension of their solution subspace is one. This subspace is spanned by $(1, 1, 0)$. In particular, the geometric multiplicity of the eigenvalue 2 is one.

Exercise 9.3

Prove that if the sum of the algebraic multiplicities of eigenvalues equals n (the dimension of the matrix), then the coefficient of x^{n-1} in the characteristic polynomial is minus the sum of the eigenvalues, where each is multiplied by its algebraic multiplicity.

Exercise 9.4

Let A be the following matrix:

$$A = \begin{pmatrix} 2 & 1 & 0 & 0 \\ 0 & 2 & 0 & 0 \\ 0 & 0 & 1 & 1 \\ 0 & 0 & -2 & 4 \end{pmatrix}.$$

(a) Compute $det(A)$.
(b) Find the eigenvalues of A.
(c) Find the algebraic and geometric multiplicities of each of the above eigenvalues.
(d) Show that the claim proved in Exercise 9.3 holds for matrix A.

Exercise 9.5

Let $A \in R^{2 \times 2}$ be a symmetric matrix, namely $A_{21} = A_{12}$.

(a) Prove the existence of the eigenvalues of A.
(b) Find an expression for the eigenvalues as a function of the matrix entries $\begin{pmatrix} a & b \\ b & d \end{pmatrix}$.
(c) Show that the two eigenvalues are different if the main diagonal entries are different.

The analysis of the eigensystem in case of a lower or upper triangular matrices is simple.

Theorem

Theorem 9.5. *Assume that $A \in R^{n \times n}$ is a lower triangular matrix. Then A_{ii} is an eigenvalue of A, $1 \leq i \leq n$. Also, the first (respectively, last) $i-1$ entries of a right (respectively, left) eigenvector associated with the eigenvalue A_{ii} are zero, $1 \leq i \leq n$. The rest of the entries can be found by forward (respectively, backward) substitutions. Finally, the algebraic and geometric multiplicities coincide across all eigenvalues.*

Proof. It is easy to possible that the characteristic polynomial of A is $\Pi_{i=1}^{n}(t - A_{ii})$ as the determinant of upper and lower triangular matrices is the product of its diagonal entries. This implies that A_{ii} is an eigenvalue of A then, immediate, $1 \leq i \leq n$. The statements regarding the corresponding eigenvectors can be checked by inspection. Moreover, the correctness of this statement implies equal multiplicities since v_i and v_{i+1} (the stated eigenvectors corresponding to the eigenvalues A_{ii} and $A_{i+1,i+1}$) are linearly independent, regardless if A_{ii} and $A_{i+1,i+1}$ are identical or not. $\square$

Note that the corresponding results for the case where A is upper triangular are immediate since then A^t is lower triangular.

Theorem

Theorem 9.6. *Let (λ_i, v_i), $1 \leq i \leq n$, be n eigenpairs of $A \in R^{n \times n}$. Assume none of the eigenvalues coincide. Then, for any scalar x,*

> **Theorem**
>
> **Theorem 9.6.** *(Continued)*
>
> 1. $(1 - t\lambda_i, v_i)$, $1 \le i \le n$, *are the* n *eigenpairs of* $I - tA$.
> 2. *In particular,* $(0, v_1)$ *and* $(1 - \lambda_i/\lambda_1, v_i)$, $2 \le i \le n$, *are the* n *eigenpairs of* $I - \lambda_1 A$.

Proof. Left as an exercise. $\qquad\square$

9.4 Similar Matrices and Their Eigensystems

This section deals with similar matrices which were defined in Section 7.3.

> **Theorem**
>
> **Theorem 9.7.** *Suppose* A *and* B *are similar (square) matrices. Then they share the same set of eigenvalues.*

We prove the theorem in two ways.

Proof. (1). We in fact show more here: A and B share the same characteristic polynomial. Indeed, as for same matrix S, $B = SAS^{-1}$, we get that

$$P_B(t) = det(tI - B) = det(tI - SAS^{-1}) = det(tSIS^{-1} - SAS^{-1})$$
$$= det(S(tI - A)S^{-1}) = det(S)det(tI - A)det(S^{-1}) = P_A(t). \qquad\square$$

Proof. (2). Since A and B are similar, then there exists a matrix S such that $B = SAS^{-1}$. Suppose $Ax = \lambda x$. Then,

$$BSx = SAS^{-1}Sx = SAx = \lambda Sx.$$

Hence, λ is an eigenvalue of B with Sx being a corresponding right eigenvector. The converse is also true as we can swap the role of A and B. $\qquad\square$

The following lemma is immediate and it will be needed for later analysis.

Lemma

Lemma 9.1.

$$det \begin{pmatrix} A & C \\ \underline{0} & B \end{pmatrix} = det(A)det(B).$$

Hence, $P_A(t)P_B(t)$ is the characteristic polynomial of the matrix on the left-hand side.

Proof of Theorem 9.4. Let r be the geometric multiplicity of the eigenvalue λ of A and let $v^1, \ldots, v^r$ be a set of r eigenvectors which form an orthonormal basis for the right eigenspace of λ. In particular, $Av^i = \lambda v^i$, $1 \leq i \leq r$. Append this set with $n - r$ vectors to form a basis for R^n. Denote by $V \in R^{n \times n}$ the matrix whose n columns are these n basis vectors. Of course, V^{-1} exists. Define the matrix $B \in R^{n \times n}$ via $B = V^{-1}AV$ which is similar to A. Then,

$$AV = VB.$$

Moreover, due to the orthonormality of the selected eigenvectors, the first r columns of B are hence λe_i, $1 \leq i \leq r$. Hence, B is a matrix of the shape dealt with in Lemma 9.1. As B is similar to A, we conclude by Theorem 9.7, that $P_A(t) = P_B(t)$. Finally, due to the shape of the matrix B and by Lemma 9.1 one can see that

$$P_B(t) = (t - \lambda)^r Q(t),$$

for some polynomial $Q(x)$ of degree $n - r$. Hence the algebraic multiplicity of λ is at least r. This completes the proof. $\square$

9.5 Bases with Eigenvectors and Diagonalizable Matrices

The following theorem states that eigenvectors of different eigenvalues need to be linearly independent.

Theorem

Theorem 9.8. *Let $\{\lambda_i\}_{i=1}^k$ be k distinguished eigenvalues with the corresponding k right (or left) eigenvectors $\{v^i\}_{i=1}^k$. Then these vectors are linearly independent.*

Proof. The proof will be by an induction argument. The case for $k = 1$ is obvious. Assume the result holds for $k - 1$. Aiming for a contradiction, assume that for some set of coefficients $\{\alpha_i\}_{i=1}^{k}$, at least two of which are not zero,

$$\underline{0} = \sum_{i=1}^{k} \alpha_i v^i. \tag{9.1}$$

Multiplying both sides by A leads to

$$\underline{0} = \sum_{i=1}^{k} \alpha_i \lambda_i v^i. \tag{9.2}$$

Multiply both sides of equality (9.1) by λ_k and subtract this equation from Eq. (9.2). You should get that

$$\underline{0} = \sum_{i=1}^{k-1} \alpha_i (\lambda_i - \lambda_k) v^i.$$

Since at least one of the coefficients on the right-hand side is not a zero, we conclude that $\{v^i\}_{i=1}^{k-1}$ are not linearly independent. This contradicts the induction hypothesis. $\qquad\square$

Note that for a set of eigenvalues which are simple, their corresponding eigenvectors are linearly independent. If, however, one of them comes with a geometric multiplicity which is larger than one, then one can look for a basis for the corresponding eigenspace. This basis coupled with the other eigenvectors form a set of linearly independent eigenvectors which span a linear subspace whose dimension is the sum of the geometric multiplicities across these eigenvalues. Of course, each time a new eigenvalue joins this set of eigenvalues, this dimension goes up by at least one. An upper bound on this sum is n. One may wish to achieve this sum but this is not always possible. As we will see in the next chapter, a sufficient condition for a success here is having a symmetric matrix.

Suppose that a basis of n eigenvectors of matrix $A \in R^{n \times n}$ exists. Denote the corresponding eigensystems by (λ_i, v^i), $1 \leq i \leq n$. Let $x \in R^n$. Thus, expressing this vector as a linear combination of these eigenvectors, we conclude that there exists a set of scalars α_i, $1 \leq i \leq n$, such that

$x = \Sigma_{i=1}^{n} \alpha_i v^i$. Then,

$$Ax = A\sum_{i=1}^{n} \alpha_i v^i = \sum_{i=1}^{n} \alpha_i A v^i = \sum_{i=1}^{n} \alpha_i \lambda_i v^i.$$

In other words, the linear function Ax is applied to each of the directions defined by v^i, $1 \le i \le n$, separately: the ith component in x is multiplied by λ_i, $1 \le i \le n$. Loosely speaking, there is no further interaction between these n directions.

We go on dealing with the special case where n simple eigenvalues $\{\lambda_i\}_{i=1}^{n}$ exist.[c] Let $\{v^i\}_{i=1}^{n}$ be a set of corresponding right eigenvectors. Denote by $D \in R^{n \times n}$ the diagonal matrix with $D_{ii} = \lambda_i$, $1 \le i \le n$. Also, let $V \in R^{n \times n}$ be a matrix whose columns are $\{v^i\}_{i=1}^{n}$, $1 \le i \le n$. Clearly, V^{-1} exits. It is easy to see that $AV = VD$ and hence

$$V^{-1}AV = D \quad \text{and} \quad A = VDV^{-1}. \tag{9.3}$$

In particular, A is similar to a diagonal matrix. A square matrix which is similar to a diagonal matrix is said to be *diagonalizable*. Finally, note the rows for V^{-1} are linearly independent right eigenvectors of A. In particular, the ith row of V^{-1} corresponds to λ_i, $1 \le i \le n$.

Example 3 (cont.). Note that for the matrix presented in Example 1,

$$A = \begin{pmatrix} 1 & 2 \\ 3 & 2 \end{pmatrix} = \frac{1}{5}\begin{pmatrix} 2 & 1 \\ 3 & -1 \end{pmatrix}\begin{pmatrix} 4 & 0 \\ 0 & -1 \end{pmatrix}\begin{pmatrix} 1 & 1 \\ 3 & -2 \end{pmatrix}. \tag{9.4}$$

We have seen above that the existence of n linearly independent eigenvectors is a sufficient condition for diagonalizability for $A \in R^{n \times n}$. Yet, it is also a necessary one. Indeed, it is possible to see that if $A = VDV^{-1}$ for some diagonal matrix D, then D_{ii} and the ith column of V form an eigensystem for A. Indeed, denote this i column of V by v^i, then,

$$Av^i = VDV^{-1}v^i = VDe_i = D_{ii}Ve_i = D_{ii}v^i,$$

---

[c]In fact, all the below holds for the case where the sum of the geometric multiplicities across the eigenvalues equals n. We assume here further that all multiplicities equal 1 in order to ease the exposition.

as required. A consequence of this observation is that not all square matrices are diagonalizable. For example, see Example 1 above which indicates that this is the case where $A \in R^{2 \times 2}$, when $A_{11}A_{22} < A_{12}A_{21}$.

Note that

$$A^k = (VDV^{-1})\cdots(VDV^{-1}) = VD^kV^{-1}, \quad k \geq 1.$$

Observe that the powers of D, as opposed to the powers of A, are easily computed. Finally, observe that λ_i^k is an eigenvalue of A^k, and with the same set of right eigenvectors, $k \geq 1$. This result can be extended to negative values of k but, of course, as long as A is regular, namely all of its eigenvalues (and not only the one under consideration) are non-zero. In particular,

---

Theorem

Theorem 9.9. *Let λ be an eigenvalue of the invertible matrix A with $Av = \lambda v$, then $A^{-1}v = \frac{1}{\lambda}v$. Moreover, their geometric multiplicities coincide.*

Proof. $Av = \lambda v$ if and only if $v = \lambda A^{-1}v$ which holds for $\lambda \neq 0$ if and only if $A^{-1}v = \frac{1}{\lambda}v$. $\qquad\square$

Note that it is not claimed that the algebraic multiplicity of λ as an eigenvalue of A coincides with the algebraic multiplicity of λ^k in A^k.

Theorem

Theorem 9.10. *Let QR be the QR factorization of a square and invertible matrix A.[a] Then the matrices A and R share the same set of eigenvalues. Moreover, their geometric multiplicities coincide.*

---
[a] *Note that the existence of Q follows the assumption that A is invertible.*

Proof. Note that if $Rv = \lambda v$, then $QRv = \lambda Qv$, namely $Av = \lambda Qv$, implying that Qv is an eigenvector of A with λ as the corresponding eigenvalue. In a similar way, note that if v is an eigenvector of A, then $Q^{-1}v$ is an eigenvector of R. $\qquad\square$

Exercise 9.6

Let

$$A = \begin{pmatrix} 1 & -3 & 3 \\ 3 & -5 & 3 \\ 6 & -6 & 4 \end{pmatrix}.$$

(a) Show that the eigenvalues of A are 4 and -2.
(b) Show that the eigenspace that corresponds to eigenvalue 4 is of dimension 1. Find the right and left eigenvectors whose scalar product is 1.
(c) Show that the eigenspace that corresponds to eigenvalue -2 is of dimension 2. Find a basis for the right eigenvectors. *Hint*: show that $x_1 = (1, 1, 0)$ and $x_2 = (1, 0, -1)$ span the eigenspace that corresponds to eigenvalue -2.
(d) Show that the right eigenvectors of eigenvalue 4 do not belong to the sub-space spanned by the basis found previously.
(e) Show that A is diagonalizable.

9.6   The Spectral Representation

Assume as in the previous section that $\{v^i\}_{i=1}^n$ is a set of right eigenvectors of A that form a basis for R^n. Denote by V the matrix whose columns are these eigenvectors. Let $\{w^i\}_{i=1}^n$ be a set of the corresponding left eigenvectors. Denote by W the matrix whose rows are these eigenvectors. Note that v^i and w^i are associated with the same eigenvalue, $1 \leq i \leq n$. Again, we do not claim that such sets always exist (and indeed in many cases they do not), so what is said below holds for the case when they do. Finally, assume further that all eigenvalues are simple.

From Theorem 9.1 we learn that $(w^i)^t v^j = 0$, $1 \leq i \neq j \leq n$. Based on Theorem 9.8, we learn that each set of vectors forms a basis for R^n. From Theorem 9.1 we also learn that WV is a diagonal matrix, with all diagonal entries being non-zero. This fact can be proved as follows: Had $(w^i)^t v^i = 0$ for some i, $1 \leq i \leq n$, then $(w^i)^t v^j = 0$ for all j, $1 \leq j \leq n$, making w^i orthogonal to all vectors in R^n. Hence, $w^i = \underline{0}$ which is a contradiction. Moreover, by the right scaling, we can assume without loss of generality that $(w^i)^t v^i = 1$, $1 \leq i \leq n$, in which case $WV = I$, or, put differently $W = V^{-1}$.

Finally, let $E_i \in R^{n \times n} = v^i(w^i)^t$, $1 \leq i \leq n$. Note that all these matrices are rank-one matrices: all their columns (and likewise all their rows) are parallel in the sense that up to a scalar multiplication, they are all identical. As an exercise, check that $E_i^k = E_i$ for any $k \geq 1$ and that $E_i E_j = \underline{0}$ for $1 \leq i \neq j \leq n$. The following theorem is known as the *spectral representation* of a matrix.

Theorem

Theorem 9.11. *Let λ_i, $1 \leq i \leq n$, be n simple eigenvalues of A. Then,*

$$A = \sum_{i=1}^{n} \lambda_i E_i \tag{9.5}$$

We prove the theorem in two ways.

Proof. (1) From (9.3) we learn that $A = VDV^{-1}$, where the V has v^i, $1 \leq i \leq n$, as its columns and where V^{-1} has w^i, $1 \leq i \leq n$, as it rows. Let $D_i \in R^{n \times n}$ be a matrix with all its entries equal zero, except for the ith diagonal entry λ_i, $1 \leq i \leq n$. Then,

$$A = VDV^{-1} = V \left(\sum_{i=1}^{n} D_i \right) V^{-1} = \sum_{i=1}^{n} VD_iV^{-1} = \sum_{i=1}^{n} \lambda_i E_i,$$

as required. $\qquad\square$

Proof. (2) First note that if for two matrices A and B, $Ax = Bx$ for any x, then $A = B$. Thus, we will show next that for any $x \in R^n$, $Ax = \sum_{i=1}^{n} \lambda_i E_i x$. Indeed, since $\{v^i\}_{i=1}^{n}$ form a basis for R^n, any $x \in R^n$ can be written as $x = \Sigma_{i=1}^{n} \alpha_i v^i$ for some scalars $\{\alpha_i\}_{i=1}^{n}$. Then,

$$Ax = A \sum_{i=1}^{n} \alpha_i v^i = \sum_{i=1}^{n} \alpha_i A v^i = \sum_{i=1}^{n} \alpha_i \lambda_i v^i. \tag{9.6}$$

On the other hand, since $E_i v^j = \underline{0}$ for any pair $1 \leq i \neq j \leq n$, and since $E_i v^i = v^i$ for $1 \leq i \leq n$, we get that

$$\sum_{i=1}^{n} \lambda_i E_i x = \sum_{i=1}^{n} \lambda_i E_i \sum_{i=1}^{n} \alpha_i v^i = \sum_{i=1}^{n} \alpha_i \lambda_i E_i v^i = \sum_{i=1}^{n} \alpha_i \lambda_i v^i,$$

which coincides with (9.6). $\qquad\square$

Since

$$E_i E_j = \begin{cases} E_i & \text{if } 1 \le i = j \le n, \\ \underline{0} & \text{if } 1 \le i \ne j \le n, \end{cases}$$

we conclude with the help of (9.5), that

$$A^k = \sum_{i=1}^{n} \lambda_i^k E_i, \quad k \ge 0. \tag{9.7}$$

In particular, the case where $k = 0$ leads to $\Sigma_{i=1}^{n} E_i = I$.

Example 2 (cont.). Note that for the matrix presented in Example 2 (see (9.4)),

$$A = \begin{pmatrix} 1 & 2 \\ 3 & 2 \end{pmatrix} = 4 \begin{pmatrix} \frac{2}{5} \\ \frac{3}{5} \end{pmatrix} \begin{pmatrix} 1 & 1 \end{pmatrix} + (-1) \begin{pmatrix} \frac{1}{5} \\ -\frac{1}{5} \end{pmatrix} \begin{pmatrix} 3 & -2 \end{pmatrix}$$

$$= 4 \begin{pmatrix} \frac{2}{5} & \frac{2}{5} \\ \frac{3}{5} & \frac{3}{5} \end{pmatrix} - 1 \begin{pmatrix} \frac{3}{5} & -\frac{2}{5} \\ -\frac{3}{5} & \frac{2}{5} \end{pmatrix}.$$

We can draw two conclusions from the Theorem 9.11. First, note that when the matrix A is applied to any vector, it operates on each of its components independently, where each component corresponds to a different eigenvector: The component of v_i, which originally came with the value of α_i is now with $\alpha_i \lambda_i$ and this is regardless of the values of the other components α_j, $j \ne i$, $1 \le i \le n$. In particular, when applying A to an eigenvector it does not mix up with other vectors: the direction stays the same if $\lambda_i > 0$ or it is reversed if $\lambda_i < 0$. Moreover, the magnitude of the change is determined by the corresponding value of λ_i. A similar result does not hold when a vector in the basis is not an eigenvector. Secondly, and assuming that $|\lambda_1| > |\lambda_2| > \cdots > |\lambda_n|$, if one replaces A with

$$\sum_{i=1}^{k} \lambda_i E_i, \tag{9.8}$$

for some $k < n$, the corresponding operation will be simpler, while the quality of the approximation depends on how small is $|\lambda_{k+1}|$ in comparison with $|\lambda_1|$.

Assume without loss of generality that λ_1 is the largest eigenvalue in absolute value, and for simplicity assume that this is the case without any ties. Then, by (9.7),

$$\lim_{k \to \infty} \frac{1}{\lambda_1^k} A^k = E_1.$$

Moreover, in the case where $\alpha_1 \neq 0$, we obtain,

$$\lim_{k \to \infty} \frac{1}{\lambda_1^k} A^k x = E_1 x = E_1 \left(\sum_{i=1}^{n} a_i v^i \right) = \alpha_1 E_1 v^1 = \alpha_1 v^1,$$

which means that in the long-run we get an eigenvector which corresponds to the largest eigenvalue, and this is regardless of the initial vector x. In the case where $\alpha_1 = 0$, the limit matrix is $\underline{0}$.

Remark. Suppose that the matrix A possesses a spectral representation $\Sigma_{l=1}^{n} \lambda_l E_l$ and it is also invertible, which is the case if and only if all its eigenvalues are different from zero. In this case,

$$A^{-1} = \sum_{l=1}^{n} \frac{1}{\lambda_l} E_l. \tag{9.9}$$

Also, when solving the system of equations $Ax = b$ by $x = A^{-1}b$, we get that

$$x = \sum_{l=1}^{n} \frac{1}{\lambda_l} E_l b.$$

How sensitive is this solution as a function of b? For this we look into the derivative of x_i with respect to b_j. It is possible to see that

$$\frac{d\,x_i}{d\,b_j} = \sum_{k=1}^{n} \frac{1}{\lambda_k} v_i^k w_j^k, \quad 1 \leq i, \ j \leq n.$$

Here the dominant factor is the smallest eigenvalue (in absolute value): the smaller it is, the more sensitive is the solution.

Remark. A related, but different, question, is when a matrix is "close" to be singular. Assume that the matrix A comes with a spectral representation. We know that when $det(A) = 0$ the matrix is singular, so a possible measure for the level of singularity is how close is $det(A)$ to zero: The closer it is, the less invertible is the matrix. This, in turn, is reflected (see (8.8)) by A^{-1} having large values. Put differently, now by (9.9), this should be reflected by the matrix A having small eigenvalues, or more precisely that $\min_i |\lambda_i|$ is small. But there is a snag here. We feel that αA should be as invertible as A is for any scalar α. However, λ is an eigenvalue of A if and only if $\alpha\lambda$ is an eigenvalue of αA. Thus, using the measure $\min_i |\lambda_i|$, we can take smaller and smaller values for α and reduce the invertability of a matrix.[d] This of course, does not make sense. Hence, we should look for a measure which is invariant with respect to α and still keeps the idea that the smaller $\min_i |\lambda_i|$ is, the less invertible is the matrix. A possible choice is

$$\frac{\max_i |\lambda_i|}{\min_i |\lambda_i|},$$

which indeed is invariant with respect to α. This value is called the *condition number*.

Exercise 9.7

Suppose that matrix A has the following spectral representation:

$$A = \sum_{i=1}^{n} \lambda_i E_i,$$

where λ_i is an eigenvalue, $E_i = x_i y_i^t$, where x_i and y_i are the corresponding right and left eigenvectors, respectively, $1 \leq i \leq n$.
(a) Prove that $\sum_{i=1}^{n} E_i = I$.
(b) Prove that for any x, which is not an eigenvalue of A, it holds that:

$$(xI - A)^{-1} = \sum_{i=1}^{n} \frac{1}{x - \lambda_i} E_i.$$

We conclude this section with the Cayley–Hamilton theorem.

Theorem

Theorem 9.12. *Any square matrix zeroes its characteristic polynomial. In other words,* $P_A(A) = \underline{0}$[a].

---

[a] *Note that $P_A(A)$ is a matrix and when one applies a polynomial on a matrix, matrix operations such as power and additions are done as defined for matrices.*

---

[d] Note $det(\alpha A)$ goes to zero even in a higher pace, that of α^n.

Proof. Let $P_A(x) = \sum_{j=0}^{n} a_j x^j$ for some coefficients a_j, $0 \leq j \leq n$. We need to show that $\sum_{j=0}^{n} a_j A^j = \underline{0}$ where $\underline{0} \in R^{n \times n}$ is a matrix having all its entries equal to zero. The theorem holds for any square matrix and there is no need to assume anything regarding its eigenvalues. In particular, there is no need to assume their existence. Yet, our proof here is done under the assumption that the matrix A comes with n simple eigenvalues. For this case

$$P_A(A) = \sum_{j=0}^{n} a_j A^j = \sum_{j=0}^{n} a_j \left(\sum_{i=1}^{n} \lambda_i E_i \right)^j = \sum_{j=0}^{n} a_j \sum_{i=1}^{n} \lambda_i^j E_i$$

$$\sum_{i=1}^{n} E_i \sum_{j=0}^{n} a_j \lambda_i^j = \sum_{i=1}^{n} E_i P_A(\lambda_i) = \sum_{i=1}^{n} E_i 0 = \underline{0}.$$

For a proof which removes the assumptions used here, see, e.g., [8, p. 308].

□

Chapter 10

Symmetric Matrices

10.1 Eigensystems and Symmetric Matrices

A square matrix A is called *symmetric* if $A^t = A$. We have encountered symmetric matrices before. Note that the matrix $A^t A$, which was introduced in Chapter 6, is symmetric. We deal here with the issue of eigensystems for such matrices. The obvious point here is that a vector is a right eigenvector if and only if it is also a left eigenvector. This, coupled with Theorem 9.1, implies that eigenvectors corresponding to different eigenvalues are orthogonal. In fact, we can say much more.

Theorem

Theorem 10.1. *For a symmetric matrix $A \in R^{n \times n}$ there exists an orthonormal basis for R^n which is formed of n eigenvectors. In particular, it is diagonalizable. Also, for any eigenvalue of A, its geometric and algebraic multiplicities coincide.*

Proof. Consider the following optimization problem.

$$x_1 \in \arg \max_{x \in R^n} x^t A x,$$

$$\text{s.t. } ||x||^2 - 1 = 0.$$

In order to solve it we use the Lagrange multipliers technique. Specifically, using (3.4) for the case where $m = 1$, we get that x_1 obeys

$$\nabla x^t A x = \lambda_1 \nabla (x^t I x - 1),$$

for some scalar λ_1. Hence,

$$2Ax_1 = 2\lambda_1 x_1.$$

In particular, λ_1 and x_1 are an eigenpair. Note that the corresponding objective function value equals $x_1^t A x_1 = \lambda_1 x_1^t x_1 = \lambda_1$. Also, x_1 may be non-unique, in which case x_1 is selected arbitrarily among all the optimizer vectors. Of course, λ_1 is unique as it equals the optimal value of the objective function.

Next for i with $2 \leq i \leq n$, define the ith optimization problem

$$x_i \in \arg \max_{x \in R^n} x^t A x,$$

$$\text{s.t. } \|x\|^2 - 1 = 0.$$

$$x_j^t x = 0, \quad 1 \leq j \leq i - 1.$$

Note that x_i is orthogonal to all vectors which solved the previous $i - 1$ optimization problems. We prove by induction that x_i, $1 \leq i \leq n$, which by definition are orthonormal, are eigenvectors of A. Moreover, the corresponding eigenvalue λ_i is the value of the ith optimal objective function value, $1 \leq i \leq n$. Also, as it turns out, λ_i is the Lagrange multiplier of the constraints $\|x\|^2 - 1 = 0$, $1 \leq i \leq n$ in the ith optimization vector. Finally, note that in case of a tie and, for example, two orthogonal vectors are optimal in the ith problem, the one who was selected is not feasible for the $(i + 1)$th problem (as it would need to be orthogonal to itself).

All was proved already for the case where $i = 1$. We next consider the ith optimization problem and assume the induction hypothesis that the result holds for all solutions up to and including $i - 1$. Solving this problem again with the Lagrange multipliers technique, we get that for some scalars λ_i and μ_{ij}, $1 \leq j \leq i - 1$,

$$2Ax_i = 2\lambda_i x_i + \sum_{j=1}^{i-1} \mu_{ij} x_j. \tag{10.1}$$

Multiplying this from the left by x_j^t, we get that

$$2x_j^t A x_i = 0 + \mu_{ij} = 0, \quad 1 \leq j \leq i - 1.$$

Invoking the induction hypothesis for $1 \leq j \leq i-1$, which says that x_j and λ_j form an eigenpair of A, leads to

$$2\lambda_j x_j^t x_i = \mu_{ij}, \quad 1 \leq j \leq i-1.$$

Since by definition x_j and x_i are orthogonal, we conclude that $\mu_{ij} = 0$, $1 \leq j \leq i-1$. Thus, by (10.1),

$$Ax_i = \lambda_i x_i,$$

as promised. From here we get that $x_i^t A x_i = \lambda x_i^t x_i$, and since $x_i^t x_i = 1$, the fact that λ_i is the objective function value follows.

Note that $\lambda_1 \geq \lambda_2 \geq \cdots \geq \lambda_n$. The fact just established that the sum of geometric multiplicities across all eigenvalues equals n, and coupled with Theorem 9.4, immediately imply that the algebraic and geometric multiplicities of an eigenvalue of a symmetric matrix coincide. $\qquad\square$

Note that the above proof was constructive in a sense that it identifies the eigenvalues and the corresponding eigenvectors as solutions of constrained optimization problems (and not only proved their existence). It stops short of telling us how to derive these values. For this one needs to resort to optimization techniques which are beyond the scope of this text.

The following is as important.

Theorem

Theorem 10.2. *Let $A \in R^{n \times n}$ be a symmetric matrix. Let $D \in R^{n \times n}$ be a diagonal matrix whose diagonal entries are the eigenvalues of A, each of which appears as many times as its multiplicy prescribes. Let $V \in R^{n \times n}$ be the matrix whose columns are orthonormal eigenvectors of A.[a] Then,*

$$A = VDV^t. \tag{10.2}$$

In particular, A is similar to a diagonal matrix.

---

[a] *Note that we preserve their order as in the corresponding eigenvalues. Moreover, if the multiplicity of an eigenvalue is larger than one, we take vectors which form a basis for their eigenspace.*

Proof. After observing that $V^t = V^{-1}$, the rest of the argument here is similar to the one which appears in Section 10.4 and will not be repeated. $\qquad\square$

Exercise 10.1

Let

$$A = \begin{pmatrix} 2 & 1 \\ 1 & 3 \end{pmatrix}.$$

(a) Find the eigenvalues of A.

(b) What is the value of the following objective? $\max_{x:\|x\|=1} x^t A x$.

(c) What is the value of the following objective? $\min_{x:\|x\|=1} x^t A x$.

(d) For each of (b) and (c) above, find the vectors x_1 and x_2 that optimize these objectives.

(e) What is the spectral representation of A?

(f) Prove that if any matrix B is symmetric, then so is B^k for any integer k (including negative).

(g) Repeat (a)–(e) for A^{-1} and A^2 (hint: use any knowledge on A instead of computing the matrices A^{-1} and A^2).

10.2 Positive Matrices

Definition

Definition 10.1. A square symmetric matrix A is called semi-positive if $x^t A x \geq 0$ for any vector x. It is called positive if $x^t A x > 0$ unless $x = \underline{0}$.[a]

---

[a]In most text the term "semi-positive" ("positive", respectively) is replaced with "semi-positive definite" ("positive definite", respectively).

Note that since for square matrix B, which is not necessarily symmetric $x^t B x = x^t (\frac{A+A^t}{2}) x$, where clearly $\frac{A+A^t}{2}$ is a symmetric matrix, the above definition can be extended to non-symmetric matrices. Yet, since the below is true only for symmetric matrices, we limit our considerations only to such matrices.

Theorem

Theorem 10.3. *A square symmetric matrix A is semi-positive (positive, respectively) if and only if all its eigenvalues are non-negative (positive, respectively).*

Proof. Suppose A is semi-positive. Let λ and v be an eigenvalue and its eigenvector, respectively. Then, $v^t A v = v^t \lambda v = \lambda ||v||^2$. As this value ought to be non-negative, we must have $\lambda \geq 0$. Conversely, let $v^1, \ldots, v^n$ be a basis for R^n formed out of eigenvectors. Let $\lambda_1, \ldots, \lambda_n$ be, in this order, the corresponding set of eigenvalues. Note that we apply here repetitions of eigenvalues if needed as specified by the corresponding multiplicity. We assume that all of them are non-negative. Express any vector $x \in R^n$ as $x = \sum_{i=1}^n \alpha_i v^i$. Then,

$$x^t A x = x^t A \sum_{i=1}^n \alpha_i v_i = x^t \sum_{i=1}^n \alpha_i \lambda_i v^i$$

$$= \sum_{i=1}^n \alpha_i (v^i)^t \sum_{i=1}^n \alpha_i \lambda_i v^i = \sum_{i=1}^n \lambda_i \alpha_i^2 ||v^i||^2,$$

which is non-negative, as required. The claim for the case where A is positive can be similarly shown. $\square$

Exercise 10.2

Consider the following symmetric matrix:

$$V = \begin{pmatrix} 1 & 0 & \rho \\ 0 & 1 & 0 \\ \rho & 0 & 1 \end{pmatrix}.$$

(a) Show that the characteristic polynomial of V equals $P_V(x) = (x-1)[(x-1)^2 - \rho^2]$.

(b) Show that the there exists three eigenvalues: 1, $1 + \rho$, and $1 - \rho$.

(c) Conclude that the matrix is positive if and only if $-1 < \rho < 1$.

Exercise 10.3

Consider the following symmetric matrix:

$$V = \begin{pmatrix} 1 & \rho & \rho \\ \rho & 1 & 0 \\ \rho & 0 & 1 \end{pmatrix}.$$

(a) Let $x = (1, -\sqrt{2}/2, -\sqrt{2}/2)^t$. Which condition needs to be imposed on ρ in order to guarantee that $x^t V x > 0$.
(b) Repeat the previous item now with $x = (1, \sqrt{2}/2, \sqrt{2}/2)^t$.
(c) What are the eigenvalues of V? Conclude that the two conditions you derived above are also sufficient for V being positive.
(d) Derive the corresponding eigenvectors and state the spectral representation of V.

Exercise 10.4

A matrix A is skew-symmetric if $A_{ij} = -A_{ji}$ for any $i \neq j$.
(a) Show that if λ is an eigenvalue of A then it is zero.
(b) Does that imply that A is singular? Prove if so, otherwise find a counter-example.

Exercise 10.5

Let $A = \begin{pmatrix} 2 & -2 \\ -2 & 5 \end{pmatrix}$ be a symmetric matrix. The matrix represents linear transformation on the elementary basis.
(a) Find a matrix P whose columns form an orthonormal basis of R^2, such that the transformation matrix is diagonal.
(b) How can you find P^{-1} in that case, without calculating the inverse matrix?
(c) What is the spectral representation of A?

Exercise 10.6

Let $A = \begin{pmatrix} a+b & b \\ b & c+b \end{pmatrix}$ be a symmetric matrix with $a, b, c \geq 0$. Prove that A is a semi-positive matrix.

Our next point is the square root of a semi-positive matrix. Let $D \in R^{n \times n}$ be a diagonal matrix with $D_{ii} = \lambda_i$, $1 \le i \le n$. Again, repetitions appear here as the multiplicities prescribe. We denote the matrix $D^{\frac{1}{2}}$ to also be a diagonal matrix, but now it is the square roots of the eignevalues which appear in the diagonal. Of course, this matrix exists if and only if $\lambda_i \ge 0$, $1 \le i \le n$, which is assumed. Moreover, $D = D^{\frac{1}{2}} D^{\frac{1}{2}}$. Let $V \in R^{n \times n}$ be the matrix whose columns are the eigenvectors of A. As said above $A = VDV^t$ and recall that V^t is also the inverse of V. Hence, it is easy to see that A itself has a square root: It is the matrix $A^{\frac{1}{2}} = VD^{\frac{1}{2}}V^t$. Indeed,

$$A^{\frac{1}{2}} A^{\frac{1}{2}} = (VD^{\frac{1}{2}}V^t)VD^{\frac{1}{2}}V^t = VDV^t = A.$$

Moreover, this square root matrix is symmetric and semi-positive too.

10.2.1 *Two criteria for the positiveness of matrices*

The first criterion is known as Sylvester's criterion. First we need the definition of the *leading minor* matrices, sometimes referred to as the *principal submatrices*. There are n such square matrices, each time taking the entries which appear in the first i rows and first i columns, $1 \le i \le n$. The first is A_{11}, the second is a two-by-two matrix

$$\begin{pmatrix} A_{11} & A_{12} \\ A_{21} & A_{22} \end{pmatrix},$$

and so on, where the nth and last one is A itself. Obviously, they are called the first, second, etc. leading minors of A.

Theorem

Theorem 10.4. *The symmetric matrix $A \in R^{n \times n}$ is positive if and only if the determinants of all its n leading minors are positive.*

Proof. [a]We start by proving sufficiency, so assume the symmetric matrix A obeys this criterion and our goal is to prove that A is positive. The proof is by induction on n. The case where $n = 1$ is clear so assume the theorem holds for $n - 1$. Next we show that A has at most one negative eigenvalue.

---

[a]This proof is based on notes by Mikhail Lavrov [6].

Aiming for a contradiction, assume it has two. Let u and v be two orthogonal eigenvectors, one for each eigenvalue. Clearly, $u^t A u < 0$ and $v^t A v < 0$. Let $w = v_n u - u_n v$. Of course, $w_n = 0$. Consider the quadratic form $w^t A w$. Since $w_n = 0$, this is in fact a quadratic form in dimension $n - 1$, involving only the $n - 1$th leading minor of A. Invoking the induction hypothesis, we conclude that $w^t A w > 0$. Also,

$$w^t A w = (v_n u - u_n v)^t A(v_n u - u_n v) = v_n^2(u^t A u) + v_n^2(v^t A v) < 0,$$

which contradicts the fact that $w^t A w > 0$. Hence, $\Pi_{i=1}^n \lambda_i$, which is the product of the n eigenvalues of A, is a product of $n - 1$ positive eigenvalues and (possibly) one negative eigenvalue. However, $\Pi_{i=1}^n \lambda_i = det(A) > 0$ (see Theorem (9.3) and note that A is one of the leading minors involved). Hence, all eigenvalues are positive and A is a positive matrix.

For the converse, assume that the symmetric matrix A is positive (which implies that the same is the case with all its leading minors). The proof will repeat the argument above. Specifically, and aiming for a contradiction, assume for some k, $1 \le k \le n$, that the first $k - 1$ leading minors come with positive determinant but the kth determinant is negative. Then, it means that this leading minor has exactly one negative eigenvalue, violating the assumption that this minor is positive.   □

The advantage of this criterion over the one stated in Theorem 10.3 is that it is easily checked without the need to resort to the computation of the eigenvalues of the matrix under consideration, the latter being involved with finding the roots of high order polynomial.

The second criterion stated next is only a sufficient one but it turns out to be useful in many applications. But first we need the following definition. We say that a matrix is *irreducible* if when we form a graph with n nodes and a directed arc from node i to node j exists if and only if $A_{ij} \ne 0$, $1 \le i, j \le n$, then there will be a directed path between any ordered pair of nodes. We state the result without a proof.

Theorem

Theorem 10.5. *An irreducible symmetric matrix $A \in R^{n \times n}$ is positive if*

$$A_{ii} \ge \sum_{j \ne i} |A_{ij}|, \quad 1 \le i \le n,$$

with at least one of these n inequalities being strict.

10.3 Covariance Matrices

In this section, we relate semi-positive matrices with covariance matrices (sometimes called variance–covariance matrices). Covariance matrices are defined as follows. Let $X_i \in R^n$, $1 \leq i \leq m$, be m vectors. Define $X \in R^{m \times n}$ as the matrix whose ith row coincides with X_i, $1 \leq i \leq m$. One can look at X_{ij} as the value of variable i, $1 \leq i \leq m$, at individual j, $1 \leq j \leq n$, in a population of size n. Using Section 7.3 terminology, $X^t \in R^{n \times m}$ is the design matrix. Define the symmetric matrix $\mathrm{Var}(X) \in R^{m \times m}$ via $\mathrm{Var}_{ij}(X) = \mathrm{Cov}(X_i, X_j)$, $1 \leq i, j \leq m$.[b] In particular, $\mathrm{Var}_{ii}(X) = \mathrm{Var}(X_i)$, $1 \leq i \leq m$. Before stating the main result of this section we need the following lemma.

> **Lemma**
>
> **Lemma 10.1.** *Let* $\mathrm{Var}(X) \in R^{m \times m}$ *be the covariance matrix of some matrix* $X \in R^{m \times n}$. *Also, let* $A \in R^{p \times m}$. *Then,*
>
> $$\mathrm{Var}(AX) = A\mathrm{Var}(X)A^t \in R^{p \times p}. \qquad (10.3)$$

Proof. For $1 \leq i, j \leq p$,

$$\mathrm{Var}_{ij}(AX) = \mathrm{Cov}((AX)_i, (AX)_j) = \mathrm{Cov}\left(\sum_{k=1}^{m} A_{ik} X_k, \sum_{l=1}^{m} A_{jl} X_l\right)$$

$$= \sum_{k=1}^{m} \sum_{l=1}^{m} A_{ik} \mathrm{Cov}(X_k, X_l) A_{jl} = (A\mathrm{Var}(X)A^t))_{ij},$$

where the last equality follows some algebra. $\square$

Note that AX above can be looked as transforming the data given in X via a linear transformation from R^m to R^p. In other words, for each of the n individuals, the raw data presented by m variables is now put in terms of p variables, each of which is a linear combination of the original m variables.

The following lemma is also required. It is based on [4].

---

[b]Recall that for $1 \leq i, j \leq m$,

$$\mathrm{Cov}(X_i, X_j) = \frac{1}{n} \sum_{k=1}^{n} (X_{ik} - \overline{X}_i)(X_{jk} - \overline{X}_j),$$

where $\overline{X}_i = \frac{1}{n}\Sigma_{j=1}^{n} X_{ij}$, namely the mean value of the entries in the vector X_i, $1 \leq i \leq m$.

> **Lemma**
>
> **Lemma 10.2.** *Let $X_i \in R^{m+1}$ be the ith unit vector in R^m with a constant c appended as its $m+1$th entry, $1 \le i \le n$. Then, there exists a constant c such that $\text{Cov}(X_i, X_j) = 0$, $1 \le i \ne j \le m$.*

Proof. For $1 \le i \ne j \le m$,

$$(m+1)\text{Cov}(X_i, X_j) = (X_i - \overline{X_i})^t (X_j - \overline{X_j}) = \left(X_i - \frac{1+c}{m+1}\mathbf{1}\right)^t$$

$$\times \left(X_j - \frac{1+c}{m+1}\mathbf{1}\right)$$

$$= X_i^t X_j - 2\frac{(1+c)^2}{m+1} + \frac{(1+c)^2}{m+1} = c^2 - \frac{(1+c)^2}{m+1}.$$

Equating this to zero, we get that c obeys $mc^2 - 2c - 1$, from which we can observe that $c = \frac{1 \pm \sqrt{1+m}}{m}$. $\qquad\square$

Referring to the above lemma, we notice that $\text{Var}(X_i)$ is identical across all values of i, $1 \le i \le m$. Moreover, as we can learn from the above proof, for $j \ne i$, $(m+1)\text{Var}(X_i) = (m+1)\text{Cov}(X_i, X_j) + 1 = 0 + 1 = 1$. Hence, $\text{Var}(X_i) = 1/(m+1)$, $1 \le i \le m$. Then, let $Z_i = \sqrt{m+1}X_i$, $1 \le i \le m$ and define $Z \in R^{m \times (m+1)}$ accordingly. Clearly, $\text{Var}(Z) = I \in R^{m \times m}$. In summary, there exists empirical multi-variable data whose covariance matrix is the identity matrix.

> **Theorem**
>
> **Theorem 10.6.** *Let $A \in R^{m \times m}$ be a symmetric matrix. Then, there exists a matrix $X \in R^{m \times n}$ for some n whose covariance matrix equals A if only if A is semi-positive.*

Proof. Assume that A is the covariance matrix of $X \in R^{m \times n}$. Let $a \in R^m$. Then by (10.3), $\text{Var}(a^t X) = a^t \text{Var}(X)a = a^t A a$. As the variance of some variable this needs to be non-negative. As this inequality holds for any vector a, we conclude that A is a non-negative matrix. For the converse assume that $A \in R^{m \times m}$ is semi-positive and let $Z \in R^{m \times m+1}$ be with $\text{Var}(Z) = I$, whose existence was just established. Define $X \in R^{m \times (m+1)}$

as $A^{\frac{1}{2}}Z$ and recall that $A^{\frac{1}{2}}$ exists (see Section 10.2) and it is symmetric. Then, by (10.3),

$$\text{Var}(X) = A^{\frac{1}{2}}\text{Var}(Z)(A^{\frac{1}{2}})^t = A^{\frac{1}{2}}I(A^{\frac{1}{2}})^t = A^{\frac{1}{2}}A^{\frac{1}{2}} = A.$$

This concludes the proof. □

10.4 Computing Eigensystems: The Power Method and Deflation

In order to compute the largest (in absolute value) eigenvalue and the corresponding eignevalue for symmetric matrices, we start with an iterative procedure, known as the *power method*. This procedure does its task under the assumption that this eigenvalue is simple.

Let v_i, $1 \leq i \leq n$, be a set of n orthonormal eigenvectors of A. Also, let $E_i = v_i v_i^t$, $1 \leq i \leq n$, be the corresponding rank-one projection matrices. Recall that $E_i^2 = E_i$, $1 \leq i \leq n$ and that $E_i E_j = \underline{0}$, $1 \leq i \neq j \leq n$. Moreover, by the spectral representation theorem (9.11), $A^k = \Sigma_{i=1}^n \lambda_i^k E_i$. Then, for any $x \in R^n$ with $x = \Sigma_{i=1}^n \alpha_i v_i$, we get that

$$A^k x = \left(\sum_{i=1}^n \lambda_i^k E_i\right)\left(\sum_{i=1}^n \alpha_i v_i\right) = \sum_{i=1}^n \alpha_i \lambda_i^k v_i$$

and that $||A^k x||^2 = \Sigma_{i=1}^n \alpha_i^2 \lambda_i^{2k}$ by the orthonormality of v_i, $1 \leq i \leq n$. Then,

$$\frac{1}{||A^k x||}A^k x = \frac{1}{\sqrt{\Sigma_{i=1}^n \alpha_i^2 \lambda_i^{2k}}}\sum_{i=1}^n \alpha_i \lambda_i^k v_i$$

$$= \frac{1}{\sqrt{\alpha_1^2 + \Sigma_{i=2}^n \alpha_i^2 \frac{\lambda_i^{2k}}{\lambda_1^{2k}}}}\left(\alpha_1 v_1 + \sum_{i=2}^n \alpha_i \frac{\lambda_i^k}{\lambda_1^k}v_i\right).$$

Assuming that $\alpha_1 \neq 0$ and that $|\lambda_1| > |\lambda_i|$, $2 \leq i \leq n$, and taking the limit when k goes to infinity, leads to (a constant multiplication of) v_1, the eigenvector associated with the largest eigenvalue. Note that the greater λ_1 is in comparison with the other eigenvalues, the faster the convergence is. Thus, the ratio between the second largest and the largest eigenvalues defines the rate of convergence of the power method.

Thus the power method provides, that as long as x has a component in the direction of v_1, convergence to (a constant multiplication of) v_1

is guaranteed. In practice, in order not to run into too large values (if $|\lambda_1| > 1$), or too small ones (if $|\lambda_1| < 1$), scaling $A^k x$ to $A^k x / \|A^k x\|$ once in a while, and re-initiating the power method from the just scaled vector, is advised. In the unlikely event that $a_1 = 0$ but $\alpha_2 \neq 0$, the convergence is to v_2. In fact, the convergence is to the first v_i whose $\alpha_i \neq 0$, $1 \leq i \leq n$.

Once v_1 (or a good approximation thereof) is at hand, finding λ_1 is immediate: divide any component of $A v_1$ by the corresponding component in v_1. Alternatively, note that $\lambda_i = \frac{1}{\|v_i\|^2} v_i^t A v_i$, $1 \leq i \leq n$.

Consider next the matrix $A - \lambda_1 E_1$ (which you now have). It is easy to see that its eigenvalues are 0 and λ_i, $2 \leq i \leq n$. In particular, its smallest (in absolute value) eigenvalue is 0. Moreover, the original eigenvectors are preserved. Thus, if the power method is applied to it, the second largest (in absolute value) eigenvalue of A, λ_2, coupled with the rank-one matrix E_2, are computed. Once this is done, move on with $A - \lambda_1 E_1 - \lambda_2 E_2$, etc. This procedure, where in each time one subtracts such a rank-one matrix, is called *deflation*.

10.5   Cholesky Factorization

Let A be a semi-positive symmetric matrix. As in (10.2), write A as $V^t D V$. Define $B = V^t S V$ where $S \in R^{n \times n}$ is a diagonal matrix with $S_{ii} = \sqrt{D_{ii}}$, $1 \leq i \leq n$. Clearly, $B^t = B$ and $B^2 = A$. In other words, $B = A^{1/2}$. Note that the matrix B is not well-defined in the case where A is not a semi-positive matrix.

Theorem

Theorem 10.7. *Let A be a semi-positive matrix and let B be a symmetric matrix with $B^2 = A$. Also, let QR be the QR factorization of B. Then, $A = R^t R$. In particular, A is the product between a lower-triangular matrix (which is R^t) and its transpose (which is R).*

Proof. Let B be a square root of A which as discussed above exists and QR be its QR factorization. Recall that Q is a unitary matrix and that R is an upper-triangular matrix. Then,

$$A = BB = B^t B = (QR)^t QR = R^t Q^t QR = R^t R. \qquad \square$$

The above theorem leads to an algorithm for finding a lower-triangular matrix L such that $A = LL^t$. This is the matrix R^t looked at in

Theorem 10.7. It is computed as follows. Firstly, derive the eigensystems of A, compute the square roots of its eigenvalues, leading to the square root B. Secondly, perform the Gram–Schmidt process on B, leading to its QR factorization, $B = QR$. The resulting R^t is the matrix L you are looking for. Note that in the case where the input matrix A is not semi-positive, the algorithm will alert that, since at least one of the eigenvalues computed on the way will turnout to be negative.

A better way exists for the case where A is positive. We next state an alternative algorithm, knows as *Cholesky factorization*, for computing R^t which detours the need to first derive the QR factorization of a square root of A and will apply it directly to the entries of matrix A. This algorithm, too, will also tell if the input matrix is positive or not. But before doing that we need the following lemma.

Lemma

Lemma 10.3. *Let $A \in n \times n$ be a symmetric matrix and express it as*

$$A = \begin{pmatrix} A_{11} & A_{21}^t \\ A_{21} & A_{22} \end{pmatrix},$$

where $A_{11} \in R^1$, $A_{21} \in R^{n-1}$ and $A_{22} \in R^{(n-1)\times(n-1)}$. Then, A is positive if and only if $A_{22} - A_{21}A_{21}^t/A_{11} \in R^{(n-1)\times(n-1)}$ is positive.

Proof. A is positive if and only if for any scalar y and a vector $x \in R^{n-1}$ where not both of them equal zero, $(y, x^t)A(y, x) > 0$. This is for example the case where $y = -x^t A_{21}/A_{11}$ (requiring that $x \neq \underline{0}$). But for this choice for y, some algebra implies that $0 < (y, x^t)A(y, x) = x^t(A_{22} - A_{21}A_{21}^t/A_{11})x$ when A is positive. Since this holds for any $x \neq \underline{0}$, the first direction of the lemma is established. For the converse, note first that $A_{11} > 0$ which is a necessary condition for A to be positive. Then, if $A_{22} - A_{21}A_{21}^t/A_{11}$ is not positive then there exists an $x \neq \underline{0} \in R^{n-1}$ such that $x^t(A_{22} - A_{21}A_{21}^t/A_{11})x \leq 0$. Then, for the corresponding choice for $y = -x^t A_{21}/A_{11}$, $(y, x^t)A(y, x) \leq 0$, implying that A is not positive. This completes our proof. $\square$

Recall that our goal is to find for the case of a positive matrix $A \in R^{n \times n}$, a lower-triangular matrix $L \in R^{n \times n}$ such that $A = LL^t$. Note that such an L can be written as

$$L = \begin{pmatrix} L_{11} & \underline{0}^t \\ L_{21} & L_{22} \end{pmatrix},$$

where L_{11} is a (positive) scalar, $L_{21} \in R^{n-1}$, $\underline{0} \in R^{n-1}$ and $L_{22} \in R^{(n-1)\times(n-1)}$. Toward our goal, we try to solve for

$$A = \begin{pmatrix} A_{11} & A_{21}^t \\ A_{21} & A_{22} \end{pmatrix} = \begin{pmatrix} L_{11} & \underline{0}^t \\ L_{21} & L_{22} \end{pmatrix} \begin{pmatrix} L_{11} & L_{21}^t \\ 0 & L_{22}^t \end{pmatrix} = LL^t,$$

namely $A_{11} = L_{11}^2$ (and hence $L_{11} = \sqrt{A_{11}}$), $A_{21}^t = L_{11}L_{21}^t$ (and hence $L_{21} = A_{21}/L_{11} = A_{21}/\sqrt{A_{11}}$) and $A_{22} = L_{21}L_{21}^t + L_{22}L_{22}^t =$ (and hence $L_{22}L_{22}^t = A_{22} - L_{21}L_{21}^t$ which in fact equals $A_{22} - A_{21}A_{21}^t/A_{11}$). The good news is that by (10.3) $L_{22}L_{22}^t$ is positive. The bad news is that we do not have (yet) L_{22}: What we have is only $L_{22}L_{22}^t$. However, and this is main idea behind Cholesky factorization, the above can be repeated to the matrix $L_{22}L_{22}^t$, etc. Note that at each step one more of the columns of the matrix L we look for is found. Finally, observe that at each step the square root of a new term is computed. The proved fact that the resulting submatrix is positive guarantees the positiveness of this term.

Example. The next is an example taken from [10] for a matrix A which is factorized to the form LL^t, where L is a lower-triangular matrix:

$$\begin{pmatrix} 25 & 15 & -5 \\ 15 & 18 & 0 \\ -5 & 0 & 11 \end{pmatrix} = \begin{pmatrix} 5 & 0 & 0 \\ 3 & 3 & 0 \\ -1 & 1 & 3 \end{pmatrix} \begin{pmatrix} 5 & 3 & -1 \\ 0 & 3 & 1 \\ 0 & 0 & 3 \end{pmatrix}.$$

In particular, we conclude that A is a positive matrix. How was this matrix computed? We start with

$$\begin{pmatrix} 25 & 15 & -5 \\ 15 & 18 & 0 \\ -5 & 0 & 11 \end{pmatrix} = \begin{pmatrix} L_{11} & 0 & 0 \\ L_{21} & L_{22} & 0 \\ L_{31} & L_{32} & L_{33} \end{pmatrix} \begin{pmatrix} L_{11} & L_{21} & L_{23} \\ 0 & L_{22} & L_{32} \\ 0 & 0 & L_{33} \end{pmatrix}.$$

Firstly, we deal with the first column of L:

$$\begin{pmatrix} 25 & 15 & -5 \\ 15 & 18 & 0 \\ -5 & 0 & 11 \end{pmatrix} = \begin{pmatrix} 5 & 0 & 0 \\ 3 & L_{22} & 0 \\ -1 & L_{32} & L_{33} \end{pmatrix} \begin{pmatrix} 5 & 3 & -1 \\ 0 & L_{22} & L_{32} \\ 0 & 0 & L_{33} \end{pmatrix}.$$

As for the second column:

$$A_{22} - L21L_{21}^t = \begin{pmatrix} 18 & 0 \\ 0 & 11 \end{pmatrix} - \begin{pmatrix} 3 \\ -1 \end{pmatrix} \begin{pmatrix} 3 & -1 \end{pmatrix} = \begin{pmatrix} L_{22} & 0 \\ L_{32} & L_{33} \end{pmatrix} \begin{pmatrix} L_{22} & L_{32} \\ 0 & L_{33} \end{pmatrix},$$

namely,

$$\begin{pmatrix} 9 & 3 \\ 3 & 10 \end{pmatrix} = \begin{pmatrix} 3 & 0 \\ 1 & L_{33} \end{pmatrix} \begin{pmatrix} 3 & 1 \\ 0 & L_{33} \end{pmatrix}.$$

Finally, for the last column we get that $10 - 1 \times 1 = L_{33}^2$ and hence $L_{33} = 3$.

Finally, suppose the input matrix for Cholesky factorization is not a positive matrix but the procedure is carried nevertheless. Of course, something should go wrong. Where will this take place? The answer lies in the fact that sooner or later the upper-left corner of the resulting square matrix (denoted above by $L_{22}L_{22}^t$) will turn out to be non-positive, making this submatrix (and hence, by Lemma 10.3, the input matrix) non-positive.

Once a lower-triangular matrix L with $A = LL^t$ is at hand it is possible to easily solve the linear system of equations $Ax = b$. The first step is to find a solution for $Ly = b$ by forward substitutions. Once y is at hand, the next step is to compute x which obeys $L^t x = y$ by backward substitutions.

Application: Engineering the covariance matrix. Let $X \in R^{m \times n}$ be a design matrix. It represents m individuals who possess values for n variables. Thus, X_{ij} is the value of variable j for the ith individual, $1 \le i \le m$, $1 \le j \le n$. Assume that X is a tall full-rank matrix. For simplicity assume that the values of the variables are centered around zero, namely $\underline{1}^t X = \underline{0}^t$. Thus, the positive matrix $X^t X$ is the covariance matrix of the data. Denote it by $S \in R^{n \times n}$. Suppose one wishes to convert the n variables by a linear function to some other n variables but now the covariance matrix will be matrix $\Sigma \in R^{n \times n}$. It is of course assumed that Σ is a symmetric non-negative matrix. Thus, we look for a matrix $A \in R^{n \times n}$ such that $(XA)^t(XA) = \Sigma$, or $A^t SA = \Sigma$. Note that A multiplies X from the right. Using the Cholesky factorization technique we can find a lower triangular matrix L such that $LL^t = \Sigma$. Hence, we look for an A such that $LL^t = A^t S^{\frac{1}{2}} S^{\frac{1}{2}} A$. Taking a matrix A such that $L = A^t S^{\frac{1}{2}}$, namely $A = S^{-\frac{1}{2}} L^t$ will do. Note that we got more than we have bargained for: the change of scales was done through an upper-triangular matrix. Note that if $S = I$, namely the original variables were uncorrelated, then A is the upper-triangular matrix resulting from applying the Cholesky factorization to the matrix Σ. Conversely, if $\Sigma = I$, namely we wish to transform the original data given in a some design matrix to that with uncorrelated variables, we get that $A = S^{-\frac{1}{2}}$. This fact can also be seen with the help of (10.3).

10.6 Principal Components Analysis (PCA)

Recall the design matrix $X \in R^{n \times p}$ we have defined in Section 7.3 on multiple linear regression analysis. Assume further that none of its columns is a constant column. We like to pose here the (not well defined question) of what is the linear combination of the columns of X which contains the most information carried by these columns (that's why columns with constants are ignored). Note that this has got nothing to do with the Y variable. For simplicity, but without loss of generality, assume all the variables are centered around zero, namely $\underline{1}^t X = \underline{0}^t$.[c] Consider various sets of coefficients $w = (w_1, \ldots, w_p)$, all scaled such that $||w|| = 1$. We like to say (in fact, to define) that one linear combination w, better put in this context as a direction defined by w, is better than another, if the corresponding n values $\Sigma_{j=1}^p w_j X_{ij}$, $1 \leq i \leq n$, come with a higher variance. Yes, this is not a mistake, the higher the variance the better, as the most information is obtained. Indeed, high variance captures the variability in X in general in the population. Thus, our goal is to find such $w \in R^p$ which leads to the highest variance. Hence and since $\Sigma_{i=1}^n \Sigma_{j=1}^p w_j X_{ij} = 0$, the optimization problem we need to solve is

$$\max_{w \in R^p} \sum_{i=1}^n \left(\sum_{j=1}^p w_j X_{ij} \right)^2 ,$$

$$\text{s.t. } ||w|| = 1.$$

Inspecting the objective function we can see that it equals $w^t X^t X w$, which is clearly non-negative for any $w \in R^p$. Note that $X^t X \in R^{p \times p}$ is the covariance matrix of X, or better put, the empirical covariance of the p variables, each of which is sampled m times (since we assume that each column is normalized such that the sum across its entries equals zero). In particular, it is a semi-positive matrix. This problem was just solved: see Theorem 10.1. The optimal objective function equals the largest eigenvalue of $X^t X$ and the all important solution itself which is the (normalized)

---

[c]If this is not the case one can replace A_{ij} with $A_{ij} - \frac{1}{n}\Sigma_{k=1}^n A_{kj}$, $1 \leq i \leq n$, $1 \leq j \leq p$, and get a matrix possessing this property.

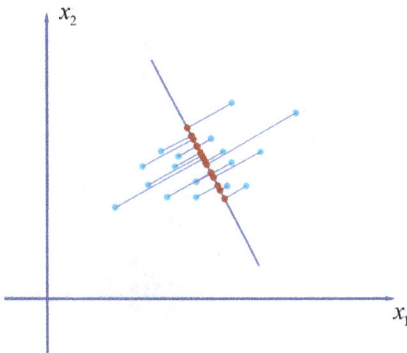

The red dots are data points projected by w_1

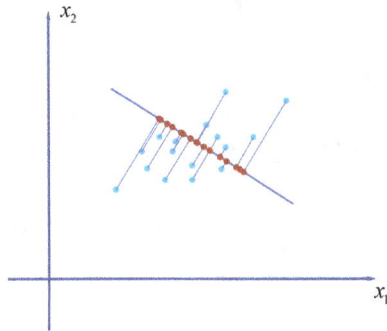

We choose another w_1 such that $w_1'X'Xw_1$ is larger.
The red dots spread out more.

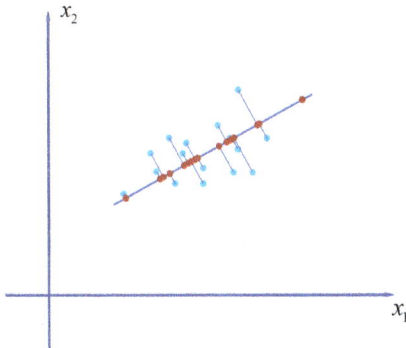

We choose the w_1 such that $w_1'X'Xw_1$ is the largest.
The red dots spread out the most.

Figure 10.1. Illustration of the PCA.

eigenvector associated with it. See Fig. 10.1 for an illustration for the case where $p = 2$.

Denote by $w^1 \in R^p$ the optimal solution and assume for simplicity that it is unique. The vector Xw^1, in fact its normalized version, $Xw^1/\|Xw^1\|$, is then referred to as the most important component of the data presented by X: much of the variability in X across the n-sized population is captured while progressing along this direction. A possible measure for quantifying this importance is

$$\frac{\lambda_1}{\sum_{i=1}^{p} \lambda_i},$$

where λ_1 is the largest eigenvalue of $X^t X$ (which is associated with w^1) and λ_i, $2 \le i \le p$, are the rest of the eigenvalues.[d] This can be explained by the fact that λ_1 is the optimal value of the objective function and it measures the achieved variability. Recall that all these eigenvalues are non-negative. Consider then the second largest eigenvalues of $X^t X$. The corresponding eigenvector leads to the second in importance direction in the data. How to derive the second (and then the third, etc.) eigenvalues is the subject of Section 10.4.

---

[d]Each eigenvalue appears as its multiplicity provides.

Chapter 11

Singular Value Decomposition

11.1 Introduction and Some Preliminaries

The issue of square matrices and diagonalization was introduced in Chapter 9 where we dealt with eigensystems. In the previous chapter we showed that symmetric matrices well-behave in the sense that they are diagonalizable. It is hence clear by now that some matrices are not diagonalizable. What can be second best? One route, which we do not take, is to introduce complex numbers. Another is to define the concept of *singular value decomposition* (SVD). Its advantage (on top of not having to use complex variables) is that it is not limited to square matrices. This is the subject of this chapter. But first we would like to establish a few results which will be useful later.

Theorem

Theorem 11.1. *Let $A \in R^{m \times n}$. Denote by $p \leq \min\{m, n\}$ the number of positive eigenvalue of $A^t A$ (repetition due to multiplicities are allowed). Let $\lambda_i > 0$ be such an eigenvalue, $1 \leq i \leq p$ and denote by v^i, $1 \leq i \leq p$, a corresponding set of orthogonal eigenvectors of $A^t A$. Then Av^i, $1 \leq i \leq p$, are non-zero orthogonal vectors.*

Proof. First, had $Av^i = \underline{0}$, then $A^t Av^i = 0$, contradicting the assumption that v^i is an eigenvector of $A^t A$ with a corresponding positive eigenvalue, $1 \leq i \leq p$. Second,

$$(Av^i)^t Av^j = (v^i)^t A^t Av^j = \lambda_j (v^i)^t v^j = \begin{cases} 0 & i \neq j \\ \lambda_j ||v^j||^2 & i = j \end{cases}, \quad 1 \leq i, j \leq p.$$
$\square$

> **Theorem**
>
> **Theorem 11.2.** *Let $A \in R^{m \times n}$ be a full-rank tall matrix and denote by v^i, $1 \leq i \leq n$, a set of n orthogonal eigenvectors of $A^t A$ (which is known to exist), then Av^i, $1 \leq i \leq n$, are all non-zero vectors and they span the image of the linear function $f : R^n \to R^m$, Ax.*

Proof. The result is immediate from the just shown fact that Av^i, $1 \leq i \leq n$, are non-zero orthogonal and hence linearly independent vectors in R^n, all in the image of f which is a linear subspace of R^m. Since $dim(img(f)) = n$, by the full rank assumption, we conclude that they form a basis for $img(f)$. $\qquad\square$

The matrix $A^t A$ was introduced in Chapter 7 and its usefulness was already established there. We next collect a few results on this matrix.

> **Theorem**
>
> **Theorem 11.3.** *Let $A^{m \times n}$. Then both $A^t A \in R^{m \times m}$ and $AA^t \in R^{n \times n}$ are symmetric semi-positive matrices.*

Proof. For any $x \in R^n$,

$$x^t(A^t A)x = (Ax)^t Ax = ||Ax||^2 \geq 0.$$

The proof for the latter part is immediate: What is true for any A, holds for A^t as well. $\qquad\square$

> **Theorem**
>
> **Theorem 11.4.** *$\lambda > 0$ is an eigenvalue of $A^t A$ if and only if it is an eigenvalue of AA^t. Moreover, their multiplicities coincide.*

Proof. Suppose $A^t Av = \lambda v$ for some non-zero vector v. Also, Av is a non-zero vector. Then $AA^t Av = \lambda Av$ or $(AA^t)Av = \lambda Av$. In other words, Av is eigenvector of AA^t with the same eigenvalue λ. Now invoke Theorem 11.1 and the fact that $A^t A$ is symmetric (making algebraic and geometric multiplicities agree). This concludes the proof. $\qquad\square$

Corollary 11.1. *Let $A \in R^{m \times n}$ be a tall matrix, namely $m \geq n$. Then the multiplicity of the eigenvalue zero of $AA^t \in R^{m \times m}$ is at least $m - n$.*

Proof. Since $A^t A \in R^{n \times n}$ and $AA^t \in R^{m \times m}$ are symmetric matrices, then by Theorem 10.2 the sum of the multiplicities of their eigenvalues are n and m, respectively. The fact that $n \geq m$ coupled with the Theorem 11.4 concludes the proof. □

Note that Theorem 11.4 holds also where $\lambda = 0$ in the case where A is a square matrix (in which case both A and A^t are singular). Note also that if for a tall matrix $A = QR$, where this is its QR factorization, then $A^t A = R^t R$. Similarly, observe that $AA^t = QRR^t Q^t$. The following is now immediate.

Corollary 11.2. *Let $A^{m \times n}$ with $m > n$ and assume it is a full-rank matrix. Then $A^t A$ is a positive matrix. Moreover, AA^t is semi-positive, but not a positive, matrix.*

Proof. The fact that both matrices are semi-positive was already shown in Theorem 11.3. Aiming for a contradiction, assume $A^t A$ is a semi-positive but not positive matrix. Hence, an $x \neq \underline{0}$ with $x^t A^t A x = ||Ax||^2 = 0$ exists. Then, $Ax = \underline{0}$, violating the assumption that A is a full-rank matrix. As for the second part, note from the previous corollary that zero is an eigenvalue of AA^t. Let $v \neq \underline{0}$ be a corresponding eigenvector. Then, $AA^t v = \underline{0}$, therefore $v^t(AA^t)v = 0$ and hence AA^t is not a positive matrix. □

11.2 Singular Value Decomposition

Assume next that $A \in R^{m \times n}$ with $m \geq n$, namely it is a square or a tall matrix. We further assume that it is a full-rank matrix. This implies that $A^t A$ is invertible and that all its eigenvalues are positive. Let $V \in R^{n \times n}$ be the matrix whose columns are the orthonormal eigenvectors of $A^t A$. Recall that $V^t V = I = VV^t$. Let $D \in R^{n \times n}$ be the diagonal matrix all its diagonal entries are the corresponding (positive) eigenvalues of $A^t A$ (in the same order). Denote by S the diagonal matrix with $S_{ii} = \sqrt{D_{ii}}$, $1 \leq i \leq n$. Note that $S^2 = D$. Clearly, S^{-1} exists and it is a diagonal matrix with $S_{ii}^{-1} = 1/S_{ii}$, $1 \leq i \leq n$. Denote S_{ii} by σ_i, $1 \leq i \leq n$, and they are called the *singular values* of A (and not of $A^t A$). Of course, they are the square root of the eigenvalues of $A^t A$. Denote the vector $\frac{1}{\sigma_i} Av^i$ by $u^i \in R^m$, $1 \leq i \leq n$. This set of vectors was in fact shown to be orthogonal in Theorem 11.1, while the normalization is easily seen. Indeed, u^i was defined this way (and not for example as Av^i) in order to have $||u^i|| = 1$, $1 \leq i \leq n$. Denote the matrix whose columns are $u^1, \ldots, u^n$ by $U \in R^{m \times n}$, then as said $U^t U = I \in R^{n \times n}$, namely U is a unitary matrix. Also, observe that $U = AVS^{-1}$. The following is now immediate.

Theorem

Theorem 11.5. *The matrices U and V are unitary matrices, with*

$$AV = US, \quad A^t U = VS, \tag{11.1}$$

and

$$A = USV^t. \tag{11.2}$$

Equation (11.2) is referred to as the SVD of the matrix A. Recall that v^i is the ith column of V and u^i is the ith column of U, $1 \leq i \leq n$, and hence we get from (11.1) that

$$Av^i = \sigma_i u^i, \quad A^t u_i = \sigma_i v^i, \ 1 \leq i \leq n.$$

The singular values of A resemble eigenvalues but they are not as they are the square roots of eigenvalues of another matrix, that of $A^t A$. Yet, in the case where A is a symmetric matrix its eigenvalues play the role of the singular values, although they are not necessarily positive as it appears in the above definition for singular values. The duality between v^i and u^i, $1 \leq i \leq n$, is also shown in the next theorem.

Theorem

Theorem 11.6. *Let u^i be the ith column of the matrix U, $1 \leq i \leq n$. Then, u^i is an eigenvector of AA^t which corresponds to the eigenvalue σ_i^2, $1 \leq i \leq n$.*

Proof. From (11.1), we learn that $A^t U = VS$. Multiplying both side from the left by A, we get that $AA^t U = AVS$. But $AV = US$, hence $AA^t U = US^2$. The facts that $S^2 = D$ and that $D_{ii} = \sigma_i^2$, $1 \leq i \leq n$, concludes the proof. $\qquad \square$

The SVD also leads to a spectral representation. Specifically, it is immediate from (11.2) that

$$A = \sum_{i=1}^{n} \sigma_i u^i (v^i)^t = \sum_{i=1}^{n} (Av^i)(v^i)^t. \tag{11.3}$$

In this way A is written as the sum of n rank-one matrices. Let $x \in R^n$. Then, for some coefficients α_i, $1 \le i \le n$, $x = \sum_{i=1}^{n} \alpha_i v^i$. Hence,

$$Ax = \sum_{i=1}^{n} \sigma_i u^i (v^i)^t \sum_{i=1}^{n} \alpha_i v^i = \sum_{i=1}^{n} \alpha_i \sigma_i u^i. \tag{11.4}$$

There is an interesting interpretation for this result. When we apply the matrix A to some vector $x = \sum_{i=1}^{n} \alpha_i v^i \in R^n$, we in fact apply it to each of its components with respect to the basis $\{v^1, \ldots, v^n\}$ individually: it replaces v^i by $u^i = \frac{1}{\sigma_i} Av^i$, which can be seen as a rotation as the norm (of 1) is preserved, while the corresponding coefficient α_i is multiplied by σ_i, $1 \le i \le n$ (which can be seen as a change of magnitude or as of scaling). This resembles the stronger result (9.6) we have when we looked into eigensystems. Note however, that this representation is trivial once we recall that $u^i = \frac{1}{\sigma_i} Av^i$, $1 \le i \le n$. This may bring one to the following (wrong) conclusion: Instead of defining u^i through the division by σ_i, one can divide it by any other value, say τ_i (making $\tau_i = 1$, $1 \le i \le n$ a natural choice), and get that $A = \sum_{i=1}^{n} \tau_i u^i (v^i)^t$. Well, this is technically correct but what will be ruined now is that u^i, $1 \le i \le n$, are not orthonormal anymore, but only orthogonal. In particular, U would not be a unitary matrix, namely a matrix which preserves the norm of a vector when multiplied with. In summary, the matrices V and U in SVD are unitary matrices and hence change the direction of the vectors they are applied to, something we refer to as rotation. On the other hand, the matrix S changes the magnitude of components of vectors when expressed as a linear combination of vectors in the basis V. This is done in the same manner as the diagonal matrix of eigenvalues behaves when a matrix is diagonalizable. Figure 11.1 depicts the SVD for the matrix

$$A = \begin{pmatrix} 0.5 & -2.5 \\ 1 & -0.5 \end{pmatrix},$$

$v^1 = (-0.2898, 0.9571)^t$, $v^2 = (0.9571, 0.2898)^t$, $\sigma_1 = 2.6514$, $\sigma_1 = 0.8486$, $u^1 = (-0.9571, -0.2898)^t$ and $u^2 = (-0.2898, 0.9571)^t$.

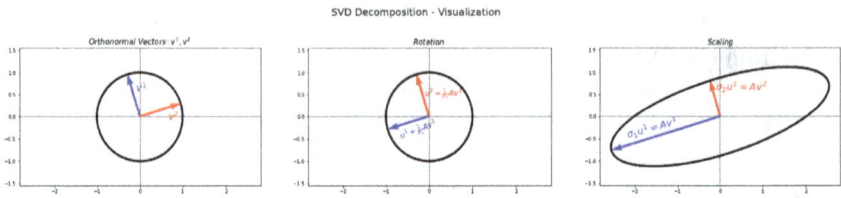

Figure 11.1. SVD visualization.

Remark. It is clear that in the case where the matrix A is wide, we can do all of the above with A^t, find the SVD of A^t, and then use the fact that $(ABC)^t = C^t B^t A^t$ in order to get the SVD of A. An alternative and a more concise way, is as follows. Back to the case where A is tall. Recall from Theorem 11.6 that the columns of U are n orthonormal eigenvectors of AA^t corresponding to the non-zero eigenvalues of AA^t. But this matrix also possesses an additional $m - n$ orthonormal eigenvectors corresponding to its zero eigenvalue. Redefine the U matrix by appending the original U with these vectors, say placed to the right of the matrix. Denote this matrix by $\tilde{U}$. This will make the U matrix a square one: $\tilde{U} \in R^{m \times m}$. Next redefine matrix S by appending it with $m - n$ zero rows. Denote this matrix by $\tilde{S}$. As opposed to the original S that was a square matrix, $\tilde{S}$ is now a tall matrix as $\tilde{S} \in R^{m \times n}$. Yet, equality (11.2) is maintained:

$$A = \tilde{U}\tilde{S}V^t.$$

Suppose now that A is a wide matrix. Now AA^t is non-singular while $A^t A$ is singular and the roles of the new V and U are swapped. In particular, U is regular, while $\tilde{V}$ comes with $n - m$ zero columns. Likewise, the new $\tilde{S}$ has $n - m$ zero columns. Hence, for a full-rank matrix $A \in R^{m \times n}$, regardless of whether A is tall or wide, A can be expressed as

$$A = USV^t,$$

with $U \in R^{m \times m}$, $S \in R^{m \times n}$ and $V \in R^{n \times n}$ and where $U^t U = I$, $V^t V = I$ and S is an enlarged diagonal matrix whose first $\min\{m, n\}$ diagonal entries are positive while the rest of $\max\{m, n\} - \min\{m, n\}$ of them are zeros. Without a proof we claim that the same is the case for any matrix which is not necessarily a full-rank matrix with the exception that some of the first $\min\{m, n\}$ entries in the diagonal of S may now be zero too.

Example.[a] Let

$$A = \begin{pmatrix} 3 & 2 \\ 2 & 3 \\ 2 & -2 \end{pmatrix}.$$

A is tall matrix in $R^{3 \times 2}$. Then,

$$A^t A = \begin{pmatrix} 17 & 8 \\ 8 & 17 \end{pmatrix} \quad \text{and} \quad AA^t = \begin{pmatrix} 13 & 12 & 2 \\ 12 & 13 & -2 \\ 2 & -2 & 8 \end{pmatrix}.$$

The eigenvalues of $A^t A$ are 25 and 9 and hence the singular values of A are $\sigma_1 = 5$ and $\sigma_2 = 3$. The corresponding orthonormal eigenvectors of $A^t A$ are

$$v^1 = \begin{pmatrix} \frac{1}{\sqrt{2}} \\ \frac{1}{\sqrt{2}} \end{pmatrix} \quad \text{and} \quad v^2 = \begin{pmatrix} \frac{1}{\sqrt{2}} \\ -\frac{1}{\sqrt{2}} \end{pmatrix},$$

respectively. Two (out of the three) orthonormal eigenvectors of AA^t are $u^1 = \frac{1}{\sigma_1} A v^1$ and $u^2 = \frac{1}{\sigma_2} A v^2$. Denote by u^3 the third one corresponding to the eigenvalue 0 of AA^t. Then,

$$u^1 = \begin{pmatrix} \frac{1}{\sqrt{2}} \\ \frac{1}{\sqrt{2}} \\ 0 \end{pmatrix} \quad u^2 = \begin{pmatrix} \frac{1}{\sqrt{18}} \\ -\frac{1}{\sqrt{18}} \\ \frac{4}{\sqrt{18}} \end{pmatrix} \quad \text{and} \quad u^3 = \begin{pmatrix} \frac{2}{3} \\ -\frac{2}{3} \\ -\frac{1}{3} \end{pmatrix}.$$

The singular value decomposition of A is hence

$$A = \begin{pmatrix} 3 & 2 \\ 2 & 3 \\ 2 & -2 \end{pmatrix} = \begin{pmatrix} \frac{1}{\sqrt{2}} & \frac{1}{\sqrt{18}} & \frac{2}{3} \\ \frac{1}{\sqrt{2}} & -\frac{1}{\sqrt{18}} & -\frac{2}{3} \\ 0 & \frac{4}{\sqrt{18}} & -\frac{1}{3} \end{pmatrix} \begin{pmatrix} 5 & 0 \\ 0 & 3 \\ 0 & 0 \end{pmatrix} \begin{pmatrix} \frac{1}{\sqrt{2}} & \frac{1}{\sqrt{2}} \\ \frac{1}{\sqrt{2}} & -\frac{1}{\sqrt{2}} \end{pmatrix}.$$

Finally, using (11.3),

$$A = 5 \begin{pmatrix} \frac{1}{\sqrt{2}} \\ \frac{1}{\sqrt{2}} \\ 0 \end{pmatrix} \left(\frac{1}{\sqrt{2}}, \frac{1}{\sqrt{2}} \right) + 3 \begin{pmatrix} \frac{1}{\sqrt{18}} \\ -\frac{1}{\sqrt{18}} \\ \frac{4}{\sqrt{18}} \end{pmatrix} \left(\frac{1}{\sqrt{2}}, -\frac{1}{\sqrt{2}} \right).$$

---

[a]Based on Jonathan Hui's notes [5].

We next show that the singular value decomposition of A leads to $A^\dagger$, the pseudo-inverse of A which was dealt with extensively in Chapter 7. Recall that $A^\dagger = (A^t A)^{-1} A^t$. Indeed,

---

Theorem

Theorem 11.7. *Let $A = USV^t$ be the SVD of the matrix A. Assume that A is full-rank. In particular, its Moore–Penrose inverse matrix exists and it is denoted by $A^\dagger$. Then*

$$A^\dagger = VS^{-1}U^t.$$

Proof. Note that since a regular matrix shares its eigenvectors with its inverse, where the eigenvalues reciprocates, the fact that $A^t AV = VS^2$ implies that $(A^t A)^{-1}V = VS^{-2}$. Now,

$$A^\dagger = (A^t A)^{-1}A^t = (A^t A)^{-1}(USV^t)^t = (A^t A)^{-1}VSU^t$$
$$= VS^{-2}SU^t = VS^{-1}U^t,$$

as required.   □

Inspecting (11.4), we can see that there are singular values which are more important than others, those with a high value. Thus, it is possible that A will be represented well only if a fraction of them are used, in fact, rounding down all the others to zero. Specifically, assuming that $\sigma_1 \geq \sigma_2 \geq \cdots \geq \sigma_n$, then if we take the first k largest singular values we may well approximate A by the rank-k matrix $\Sigma_{i=1}^k \sigma_i u^i (v^i)^t$. This is the counterpart expression for the one given in (9.8) for symmetric matrices.

Application: Storing images. Suppose a black and white photo is presented by $m \times n$ pixels. It can be stored in a matrix $A \in R^{m \times n}$, where the entry A_{ij} states, in numbers, the level of blackness of the (i, j)th pixel, $1 \leq i \leq m$, $1 \leq j \leq n$. This of course requires a storage of size $m \times n$. Much storage space can be saved if SVD is preformed. In particular, if only k singular values of A are used, then the amount of storage needed is $k(1 + m + n)$, which is quite a substantial saving. Clearly, there is a trade-off between k and the quality of the stored photo. The same can be done with colored photos. Now, one requires three matrices, one for each of the primary RGB colors, red, green and blue, in order to store a picture. Let these matrices be R, G and B, all three matrices being in $R^{m \times n}$. Thus, one needs to perform SVD for each of these matrices separately, and not necessarily the same number of singular values will be utilized in each of these three matrices.

Chapter 12

Stochastic Matrices

12.1 Introduction and the Case of Positive Transition Probabilities

A square matrix $P \in R^{n \times n}$ is said to be *stochastic* (sometimes referred to as right-stochastic) if $P_{ij} \geq 0$, $1 \leq i, j \leq n$ and $\Sigma_{j=1}^{n} P_{ij} = 1$, $1 \leq i \leq n$. It is immediate that the scalar 1 and the vector $\underline{1} \in R^n$ form a right eigensystem for this matrix.

Stochastic matrices are useful in modeling a situation in which a state process evolves throughout a time period. Specifically, if at a given (discrete) time epoch the current state of nature is i, then the next one will be state j with probability P_{ij}, $1 \leq i, j \leq n$, and this is independent of the previously visited states.[a] Moreover, these transition probabilities are not a function of the time epoch the process considered is currently in. In other words, any possible state is associated with a vector of probabilities, called the *transition probabilities* which describe the statistic dynamics of the process of movement from one state to another. Such processes are called *time-homogeneous Markov chains*. Figure 12.1 gives an example for a five-state process. Note that a lack of an arc emanating in i and ending in j corresponds to zero transition probability. The corresponding

---

[a]This by no means implies that the past and the future are independent. What we have here is that given the present, they are independent.

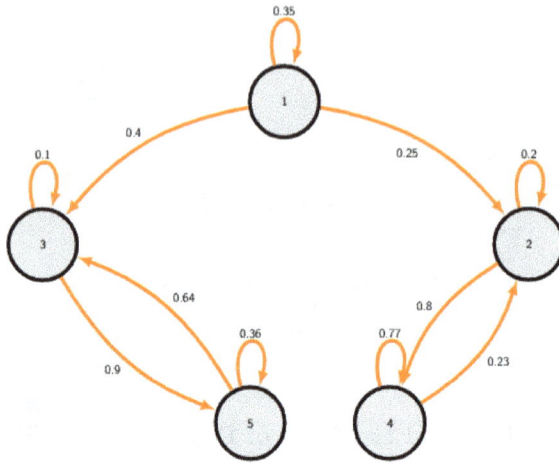

Figure 12.1. A five-state transition matrix.

transition matrix is

$$P = \begin{pmatrix} 0.35 & 0.25 & 0.4 & 0 & 0 \\ 0 & 0.20 & 0.80 & 0 & 0 \\ 0 & 0 & 0.10 & 0.9 & 0 \\ 0 & 0.23 & 0 & 0.77 & 0 \\ 0 & 0 & 0.64 & 0 & 0.36 \end{pmatrix}.$$

The first fact we like to mention is the following:

Theorem

Theorem 12.1. *If P is a stochastic matrix, then the same is the case with P^m, for any $m \geq 0$. Moreover, P_{ij}^m is the probability that a process which is currently at state i, visits state j (not necessarily for the first time) m units of time afterwards.*

Proof. The case where $m = 0$ is trivial. For $m \geq 1$, we will use an induction argument. For the case where $m = 1$, this statement is true by definition. Suppose it holds for some m. Then,

$$P_{ij}^{m+1} = (P^m P)_{ij} = \sum_{k=1}^{n} P_{ik}^m P_{kj}. \tag{12.1}$$

Using the induction hypothesis, P_{ik}^m is the probability of visiting state k in time-epoch m, $1 \leq k \leq n$. The use of the law of total probability tells us that what we have in (12.1) is the corresponding probability for visiting state j at time-epoch $m + 1$. This probability is of course non-negative. Finally, summing up these probabilities with respect to j leads to the promised value of 1. $\qquad \square$

In the rest of this section we limit our discussion to the case where $P_{ij} > 0$ for all pairs of i and j, $1 \leq i, j \leq n$. This seems to be a grave simplification, maybe leading to unrealistic modelling. Yet, in the last section we will show that in fact much can be learned and projected on the general case, once this case is analyzed completely.

Example. For some p and q, $0 < p, q < 1$, let the stochastic matrix $P \in R^{2 \times 2}$ be:

$$P = \begin{pmatrix} 1 - p & p \\ q & 1 - q \end{pmatrix}.$$

Observe that (q, p) is a left eigenvector associated with the eignevalue 1.

It is possible to prove by an induction argument that for $n \geq 1$,

$$P^n = \frac{1}{p+q} \begin{pmatrix} q & p \\ q & p \end{pmatrix} + \frac{(1-p-q)^n}{p+q} \begin{pmatrix} p & -p \\ -q & q \end{pmatrix}.$$

Note (which eases the derivation) that the product between the above two matrices equals $\underline{0} \in R^{2 \times 2}$. It is also possible to verify that $1 - p - q$ is the second eigenvalue of P with left eigenvector $(1, -1)$ and right eigenvector $(p, q)^t$. Observe that $|1 - p - q| \leq 1$, with equality if and only if $p = q = 0$ or $p = q = 1$, two options which violate our assumption that all entries of P are positive. In fact,

$$P = 1 \begin{pmatrix} \frac{q}{p+q} & \frac{p}{p+q} \\ \frac{q}{p+q} & \frac{p}{p+q} \end{pmatrix} + (1 - p - q) \begin{pmatrix} \frac{p}{p+q} & -\frac{p}{p+q} \\ -\frac{q}{p+q} & \frac{q}{p+q} \end{pmatrix}$$

is the spectral representation of P. Finally, note that unless $|1 - p - q| = 1$,

$$\lim_{n \to \infty} P^n = \begin{pmatrix} \frac{q}{p+q} & \frac{p}{p+q} \\ \frac{q}{p+q} & \frac{q}{p+q} \end{pmatrix}.$$

Note that this limit matrix is such a matrix that all its rows are identical and equal to a probability vector, which is a right-eigenvector, which in turn is associated with the eigenvalue 1.

12.2   The Limit Probabilities

We next state a revealing lemma. It says that the range of the entries for any given column of $P, P^2, \ldots, P^m, \ldots$ shrinks by a factor of at most $1 - 2\epsilon$ each time one moves along this sequence, where $\epsilon = \min_{1 \leq i,j \leq n} P_{ij}$. Clearly, for $P \in R^{n \times n}$, $0 < \epsilon \leq 1/n$.

> ### Lemma
>
> **Lemma 12.1.** *For $m \geq 1$, denote $\min_i P_{ij}^m$ by m_j^m and $\max_i P_{ij}^m$ by M_j^m. Then,*
> $$M_j^{m+1} - m_j^{m+1} \leq (1 - 2\epsilon)(M_j^m - m_j^m), \quad 1 \leq j \leq n.$$

Proof. Take some vector $x \in R^n$ (in our context it is a column of P^m) and apply to it a number of weighted averages, each time with another set of weights. This is what happens n times when the matrix P multiplies x as each row of P states some positive weights which sum-up to 1. A new vector $y = Px \in R^n$ is attained. The lemma says that the range of y shrinks by a factor of at least $1 - 2\epsilon$ in comparison with that of x. Indeed, let $x_{\max} = \max_i x_i$ and $x_{\min} = \min_i x_i$ and define $y_{\max}$ and $y_{\min}$ accordingly. Then,

$$y_{\max} \leq x_{\max} - \epsilon x_{\max} + \epsilon x_{\min} = x_{\max} - \epsilon(x_{\max} - x_{\min}).$$

This is true since considering the largest number point in x, note that it swaps at least a weight of ϵ with the smallest value, leading for further reduction in its maximal value. Following the same reasoning, we get

$$y_{\min} \geq x_{\min} + \epsilon(x_{\max} - x_{\min}).$$

Subtracting the two inequalities, we conclude that

$$y_{\max} - y_{\min} \leq (1 - 2\epsilon)(x_{\max} - x_{\min}).$$

$\square$

A few things can be learned from the above theorem. First, as the range in each of the columns of P^m converges to zero, we learn that $\lim_{m \to \infty} P^m$ exists and that it is a matrix with all its rows being equal. Denote this limit matrix by P^* and its common row by π^t. Moreover, since all entries in P^m, $m \geq 1$, are positive and bounded from below by ϵ, then the same is the case with P^*. This implies that $\pi \in R^n$ is a probability vector with positive entries. Also,

$$P^* = \underline{1}\pi^t,$$

which is a rank one matrix. In summary,

> **Theorem**
>
> **Theorem 12.2.**
>
> $$\lim_{m \to \infty} P_{ij}^m = \pi_j, \quad 1 \le i, j \le n.$$
>
> *In particular, the limit probabilities are not a function of the initial state, but only of the target state.*

The terminology used here is that in the long-run, the effect (on the probabilities) of the initial states is *scrambled*.

It was already observed above that 1 is an eigenvalue of P, with a corresponding right eigenvector $\underline{1}$. The fact that $\lim_{m \to \infty} P^{m+1} = P^*$, implies that

$$PP^* = P^* = P^*P.$$

From here we can deduce that

$$\pi^t P = \pi^t,$$

namely π is a corresponding left eigenvector of P with the eigenvalue 1. Moreover, for any $m \ge 1$,

$$\pi^t P^m = \pi^t,$$

namely if the initial state is drawn with the lottery prescribed by the probability vector π, then this probability vector holds for any other time epoch (looked at as a snapshot). For this reason the limit probabilities are also called the *stationary distribution*.

Example (cont.). Recall that since $\lim_{n \to \infty}(1-p-q)^n = 0$, we concluded that

$$P^* = \lim_{n \to \infty} P^n = \frac{1}{p+q}\begin{pmatrix} q & p \\ q & p \end{pmatrix}.$$

In particular, $\pi^t = \left(\frac{q}{p+q}, \frac{p}{p+q}\right)$.

Theorem

Theorem 12.3. *The geometric multiplicity of the eigenvalue 1 of P equals 1.*

Proof. Let $u \in R^n$ be a left eigenvector of P associated with the eigenvalue 1. Then, $u^t P = u^t$. It is easy to see that $u^t P^2 = u^t P = u^t$, from which we learn that $u^t P^* = u^t$, or that $u^t \mathbf{1} \pi^t = u^t$. This is equivalent to u being some constant multiplication of π. Note that this constant equals $u^t \mathbf{1}$. □

As it turns out, one is also the algebraic multiplicity on the eigenvalue 1. We next state it below without a proof. The interested reader can find a proof in, e.g., [9, p. 7].[b]

Theorem

Theorem 12.4. *The algebraic multiplicity of the eigenvalue 1 of P equals 1.*

For completeness, note that, by Theorem 9.4, Theorem 12.3 follows from Theorem 12.4. The following result is now clear.

Theorem

Theorem 12.5.

$$rank(I - P) = n - 1.$$

Proof. Theorem 12.3 says in fact that $dim(null(I - P)) = 1$. This coupled with Theorem 5.3, implies that $rank(I - P) = n - 1$. □

Remark. π_j is the limit probability of visiting state j, $1 \leq j \leq n$. Using this definition and conditioning on the previous visited state, it also equals

---

[b]This is part of Perron–Frubenius theorem which is applicable for any matrix that obeys all the assumptions we made on P where the requirement that all row-sums equal 1 is not a necessary one. See, e.g., [9, pp. 3–7], for more on this theorem.

$\Sigma_{i=1}^{n} \pi_i P_{ij}$. This is the idea behind calling

$$x_j = \sum_{i=1}^{n} x_i P_{ij}, \quad 1 \le j \le n,$$

the balance equations.

12.2.1 *Computing the limit probabilities*

From the previous theorem we learn that the system of equations $x^t(I - P) = \underline{0}^t$ (usually called the *balance equations*) and $x^t \underline{1} = 1$ have a unique solution which is the vector π we are after. This is a set of $n + 1$ equations with n unknowns, for which we are guaranteed that a solution, $\pi \in R^n$, exists and that it is the unique one. Can it be reduced into an equivalent set of n equations? The answer is yes: out of the $n + 1$ equations, one and only one, given the others, is redundant. Which one is it? It is clearly not the equation $x^t \underline{1} = 1$ as $x^t(I - P) = \underline{0}^t$ for $x = \alpha\pi$ for any value for α. So it needs to be one of the equations $x^t(I - P) = \underline{0}^t$. In other words, one of the columns of $I - P$ needs to be replaced with a column of ones in order to get an invertible matrix. But which column is this? The answer is simple: any column. This is equivalent to saying that all sets of $n - 1$ out of the n columns of P span the same linear subspace. We will argue for that soon. Finally, replace one of the columns of $I - P$, say the first column, by a column of ones. Denote this square and invertible matrix by A. Clearly, $\pi^t A = e_1^t$, where e_1 is the first unit vector. Then, $\pi^t = e_1^t A^{-1}$, namely π^t equals the first row of the matrix A^{-1}. The fact that any column can serve our purpose follows from the following lemma.

> **Lemma**
>
> **Lemma 12.2.** *Let $x \in R^n$. Assume that $x_j = \Sigma_{i=1}^{n} x_i P_{ij}$, $1 \le j \le n - 1$, and that $x^t \underline{1} = 1$. Then, $x_n = \Sigma_{i=1}^{n} x_i P_{in}$.*

Proof.

$$x_n = 1 - \sum_{j=1}^{n-1} x_j = 1 - \sum_{j=1}^{n-1} \sum_{i=1}^{n} x_i P_{ij} = 1 - \sum_{i=1}^{n} x_i \sum_{j=1}^{n-1} P_{ij} = 1 - \sum_{i=1}^{n} x_i (1 - P_{in}),$$

where the last equality follows from the fact that P is stochastic, and moving on, we get

$$= 1 - \sum_{i=1}^{n} x_i + \sum_{i=1}^{n} x_i P_{in} = 1 - 1 + \sum_{i=1}^{n} x_i P_{in} = \sum_{i=1}^{n} x_i P_{in},$$

as required. Finally, observe that the choice of the last state, state n, was without loss of generality. □

Example (cont.). The balance equations are

$$\pi_1 = \pi_1(1 - p) + \pi_2 q,$$

and

$$\pi_2 = \pi_1 p + \pi_2(1 - q).$$

Coupled with the equation $\pi_1 + \pi_2 = 1$, they are uniquely solved by $\pi_1 = q/(p + q)$ and $\pi_2 = p/(p + q)$.

Exercise 12.1

1. Show that if $x \in R^n$, obeys $x_i P_{ij} = x_j P_{ji}$, $1 \le i, j \le n$ (called the *detailed balance equations*), then it also solves the balance equations.
2. Give an example where the detailed balance equations do not have a solution.
3. Show that if and only for any triple of nodes i, j and k, $1 \le i, j, k \le n$, $P_{ij} P_{jk} P_{ki} = P_{ik} P_{kj} P_{ji}$, then the detailed balance equations have a (unique) solution.

Exercise 12.2

Let $(A, \overline{A})$ be a partition of the state space. Show that

$$\sum_{i \in A} x_i \sum_{j \in \overline{A}} P_{ij} = \sum_{i \in \overline{A}} x_i \sum_{j \in A} P_{ij},$$

for any partition if and only if x is a constant multiplier of π.

12.3   The Deviation Matrix

Before moving on consider the following lemma:

> **Lemma**
>
> **Lemma 12.3.** *Let T be a square matrix with the property that* $\lim_{m \to \infty} T^m = \underline{0}$, *where the right-hand side is a matrix with all its entries being zero. Then $I - T$ is invertible. Moreover,*
>
> $$(I - T)^{-1} = \sum_{m=0}^{\infty} T^m. \tag{12.2}$$
>
> *In particular, the summation above converges to a finite limit.*

Proof. First note that for $m \geq 1$,

$$(I - T)(I + T + \cdots + T^{m-1}) = I - T^m. \tag{12.3}$$

Consider a row vector x such that $x^t(I - T) = \underline{0}^t$. Recall that our goal is to show that $x = \underline{0}$. Premultiply (12.3) by x and get that $\underline{0}^t = x^t - x^t T^m$ or $x^t = x^t T^m$ for all $m \geq 0$. Take limits while recalling that $\lim_{m \to \infty} T^m = \underline{0}$, and deduce that $x = \underline{0}$ and hence $I - T$ is indeed invertible. Premultiply both sides of (12.3) with $(I-T)^{-1}$, whose existence we have just established, and obtain that

$$I + T + \cdots + T^{m-1} = (I - T)^{-1} - (I - T)^{-1}T^m.$$

Take limits again. The right-hand side converges to $(I - T)^{-1}$ since T^m goes to the zero matrix. Hence, the left-hand side has a limit too, which by definition is $\Sigma_{m=0}^{\infty}T^m$. □

A matrix T with $\lim_{m \to \infty} T^m = \underline{0}$ is called a *transient matrix*. Hence, (12.2) can be looked at as a generalization of a well-known result in the case where T is a scalar whose absolute value is smaller than one. Indeed, it determines a geometric series which initiates at 1 and its common multiplication factor equals T. It is well-known that in this case, the sum of the infinite series equals $1/(1 - T)$. Also, from (12.2) we can deduce that if all the entries in T are non-negative, then the same is the case with $(I - T)^{-1}$.

The results stated next are somewhat technical. Their usefulness will be seen later. First, denote $P - P^*$ by S.

Lemma

Lemma 12.4.

- $$S^m = P^m - P^*, \quad m \geq 1. \tag{12.4}$$

  In particular, S is a transient matrix then

- $$\sum_{m=0}^{\infty} S^m = (I - S)^{-1}. \tag{12.5}$$

  In particular, the limit underlying the left-hand side exists and likewise for the inverse of $I - S$ (two properties which are not a priori guaranteed).

- $$\sum_{k=0}^{m-1} S^k = (I - S^m)(I - S)^{-1}, \quad m \geq 1. \tag{12.6}$$

Proof. Equation (12.4) can be proved by induction. Indeed, in the case where $m = 1$ the result is trivial. Then, by invoking the induction hypothesis on m,

$$S^{m+1} = S^m(P - P^*) = (P^m - P^*)(P - P^*) = P^{m+1} - P^*, \quad m \geq 0.$$

This is the case since $P^m P^* = P^* P^m = P^* P^* = P^*$. Note that (12.4) does not hold for $m = 0$ as $S^0 = I$ (and not $I - P^*$). The other two properties are based on the fact that S is transient, namely $\lim_{m \to \infty} S^m = \underline{0}$. See Lemma 12.3. $\qquad \square$

Denote by Y the matrix $(I - S)^{-1} - P^*$ or, equivalently, $(I - P + P^*)^{-1} - P^*$, which for reasons to be given shortly, is called the *deviation matrix* of P. Note that by Lemma 12.4,

$$\sum_{m=0}^{\infty} (P^m - P^*) = I - P^* + \sum_{m=1}^{\infty} (P^m - P^*) = I - P^* + \sum_{m=1}^{\infty} (P - P^*)^m$$

$$= \sum_{m=0}^{\infty} (P - P^*)^m - P^* = (I - P + P^*)^{-1} - P^* = Y. \tag{12.7}$$

It is an easy exercise to check that

$$P^*Y = \underline{0} = YP^*$$

and then that

$$Y\underline{1} = \underline{0} \quad \text{and} \quad \pi^t Y = \underline{0}^t.$$

Also,

$$Y(I - P) = Y(I - P + P^*) = [(I - P + P^*)^{-1} - P^*](I - P - P^*) = I - P^*,$$

and likewise it can be argued that

$$(I - P)Y = I - P^*. \tag{12.8}$$

In summary, the deviation matrix Y obeys

$$Y(I - P) = I - P^* = (I - P)Y. \tag{12.9}$$

One can say that Y is a type of a generalized inverse matrix of the matrix $I - P$ which does not have an inverse. Indeed, $Y(I - P)$ "almost" equals I: It deviates from I by a rank-one matrix.

The following is our main result for this section.

Theorem

Theorem 12.6. *For $m \geq 1$,*

$$\sum_{k=0}^{m-1} P^k = mP^* + Y - P^m Y. \tag{12.10}$$

Proof. The proof will be done by an induction argument. The case where $m = 1$ is obvious and follows (12.8). Assuming the condition holds for m, then in order to prove that it also holds for $m + 1$, we will show an identical property which is that

$$P^m = P^* + P^m Y - P^{m+1} Y.$$

Indeed, by (12.7)

$$P^m Y - P^{m+1} Y = \sum_{k=m}^{\infty} (P^k - P^*) - \sum_{k=m+1}^{\infty} (P^k - P^*) = P^m - P^*,$$

as required. $\qquad \square$

Remark 12.1. The left-hand side of (12.10) is a (matrix) term which goes to infinity with m. Also, as $P^*Y = \underline{0}$, the third term on the right-hand side there goes to zero when m goes to infinity. Finally, the other two terms on the right-hand side of (12.10) can be looked at as an affine function of m where P^* is the slope and Y is the intercept (in a matrix form). Note that the slope is a function only of the second coordinate defining an entry in the matrix P^* i.e., the column. This is the case since all rows of P^* are identical. In fact, the various slopes coincide with the limit probabilities, π_j, $1 \le j \le n$. In summary, $mP^* + Y$ is the asymptote of $\sum_{k=0}^{m-1} P^k$.

Remark 12.2. Inspect (12.10) again. Clearly, both $\sum_{k=0}^{m-1} P^k$ and mP^* go to infinity in all their entries when m goes to infinity. What about the difference between them, namely the limit of $\sum_{k=0}^{m-1} P^k - mP^*$? The answer is not a priori clear. In particular, there is no guarantee that a limit exists here at all and if it exists then that it is finite. What (12.10) says is that since $\lim_{m \to \infty} P^m Y = \underline{0}$, then

$$\lim_{m \to \infty} \left[\sum_{k=0}^{m-1} P^k - mP^* \right] = \lim_{m \to \infty} \sum_{k=0}^{m-1} [P^k - P^*] = Y.$$

We next interpret this equality. First, note that at time $k \ge 0$ the number of visits in state j can be either zero or one. This is known as a Bernoulli random variable. Hence, the expected number of such visits equals the probability that this variable equals 1. Given that the initial state is i, this probability equals P_{ij}^k. Thus, $\sum_{k=0}^{m-1} P_{ij}^k$ is the expected number of such visits during a horizon of length m, under these initial conditions. Next, mP_{ij}^* has the same interpretation but where the initial state is randomly selected in accordance with the probabilities π. What we have just seen is that the difference between these two terms has a limit, it is finite and equals Y_{ij}, $1 \le i, j \le n$. This is the idea behind the terminology of the deviation matrix. Note that the ratio between the two terms discussed here goes to 1 when m goes to infinity. Finally, note that $Y_{ij} - Y_{kj}$ equals the limit difference between the total number of visits in state j under two scenarios, first when i is the initial state and second, where it is state k.

Remark 12.3. As $(I - P)\underline{1} = \underline{0}$, it is clear that the square matrix $I - P \in R^{n \times n}$ is not invertible. If we "insist" on inverting this matrix we will need to delete as many rows (as many columns, respectively) as needed and get a full-rank matrix which has right (left, respectively) inverse (see

Chapter 6). A special case will be to find the Moore–Penrose inverse of the resulting non-square matrix (see Chapter 7). This is certainly a possibility and our discussion on finding π implies that in fact the rank of $I - P$ equals $n - 1$ and that any row (column, respectively) can be deleted in serving this purpose. Yet, there is a branch of linear algebra, which goes under the umbrella of *generalized inverses* where the approach is different. In particular, a generalized inverse is a square matrix in $R^{n \times n}$, namely it is of the same size as the original matrix. We will not attempt describing this vast field here but will convey some of its main ideas through the relationship between $I - P$ and Y.

Recall that $I - P$ is not invertible but inspecting (12.9) we can see that it is "almost" invertible with the "almost" inverse Y: $(I - P)Y$ which does not equal I of course, but it equals I up to some rank-one matrix. Moreover, you may recall that if $Ax = \lambda x$ where $\lambda \neq 0$, then (assuming A^{-1} exists) $A^{-1}x = \frac{1}{\lambda}x$. We can see that something of this flavor is kept when we consider $I - P$ and its generalized inverse Y. First, note that $I - P$ and Y share the same eigenvalue of zero, and with the same right and left eigenvectors being $\underline{1}$ and π^t, respectively. Moreover,

---

Theorem

Theorem 12.7. $(I - P)x = \lambda x$, where $\lambda \neq 0$ *is an eigenvalue and* x *is a corresponding right eigenvector of* $I - P$, *if and only if* $Yx = \lambda^{-1}x$. *A similar result holds for left eigenvectors.*

Proof. Let $\lambda \neq 0$ and $x \in R^n$ be such that $(I - P)x = \lambda x$. Note that x and π are orthogonal (see Theorem 9.1). Moreover, $P^*x = \underline{1}\pi^t x = 0\underline{1} = \underline{0}$. Then, $Y(I - P)x = \lambda Yx$, which by (12.9), implies that $(I - P^*)x = \lambda Yx$. Since $P^*x = \underline{0}$, we have that $x = \lambda Yx$, or $Yx = \frac{1}{\lambda}x$, as required. The converse is shown in a similar way. □

Example (cont.). Since 1 and $1 - p - q$ are the eigenvalues of P, then $1 - 1 = 0$ and $1 - (1 - p - q) = p + q$ are the eigenvalues of $I - P$. See (9.6). Moreover, the eigenvectors are preserved. Hence, the eigenvalues of Y are 0 and $1/(p+q)$, and again the eigenvectors are preserved. Hence, the spectral representation of Y is

$$Y = 0 \begin{pmatrix} q & p \\ q & p \end{pmatrix} + \frac{1}{(p+q)^2} \begin{pmatrix} p & -p \\ -q & q \end{pmatrix} = \frac{1}{(p+q)^2} \begin{pmatrix} p & -p \\ -q & q \end{pmatrix}.$$

12.4   Generalizations

Recall that we have assumed so far that all entries in the stochastic matrix P are positive. In this section we gradually relax this assumption. But first we would like to introduce the directed graph which is associated with P. It is a graph with n nodes, each being associated with one of the states underlining the process. Moreover, an arc which emanates from node i and directs to node j exists if and only if $P_{ij} > 0$. A *path* is a sequence of nodes wherein each one of them is connected by a directed arc to the next. The length of the path is the number of arcs along it. A special case of a path is a *cycle*, namely a path from a node to itself without any intermediate nodes being visited more than once. The *period* of a node is defined as the greatest common divisor of the lengths of all cycles initiating (and terminating) from that node.

12.4.1   *The irreducible and a-periodic case*

Above we have assumed that $P_{ij} > 0$ for all pair of states i and j. The following condition somewhat relaxes this assumption.

- **Communicating:** between any ordered pair of nodes i and j, $1 \leq i, j \leq n$, there exists a directed path from node i to node j.

It is possible to observe that the period of state (i.e., node) i is the greatest common divisor among all integers k with $P_{ii}^k > 0$. Indeed, whether or not P_{ij}^k is zero or not is a question which is determined only by the topology of the network associated with P, and not by the actual values of its entries. We claim, without a proof, that in a communicating process, the periodicity is a process (as opposed to a node) dependent entity, namely all periods across its nodes are equal. Put differently, the period is an equivalence property among nodes in a communicating process. For a proof, see, e.g., [1 p. 226] or [9, p. 17].

Example. For the stochastic matrix P, where

$$P = \begin{pmatrix} 0 & 1 \\ 1 & 0 \end{pmatrix}$$

the period equals 2.

For the matrix

$$S = \begin{pmatrix} 0 & 1 & 0 & 0 & 0 \\ 1 & 0 & 0 & 0 & 0 \\ 0 & 0 & 0 & 1 & 0 \\ 0 & 0 & 0 & 0 & 1 \\ 0 & 0 & 1 & 0 & 0 \end{pmatrix}$$

compute S^{150} and S^{1001} (*Hint*: find the periodicity).

The following assumption will be required for this section.

- **A-periodicity:** A communicating process (and hence its corresponding transition (stochastic) matrix) is said to be *a-periodic* if the (common) period equals one. Note that a sufficient, but not a necessary, condition for a-periodicity is $P_{ii} > 0$ for some node i, $1 \le i \le n$.

We claim that if the above two conditions hold, then for some m large enough, $P_{ij}^{m+k} > 0$ for any $k \ge 0$ and $1 \le i, j \le n$. We do not prove this claim but would like to say that the argument needed here is based on number theory (and not on probability theory). This is the case due to the fact that once $P_{ij} > 0$, its actual value is immaterial when the issue of a-periodicity is introduced. For a proof, see, e.g., [9, pp. 18–19]. Note that if $x^t P = x^t$ then for any $m \ge 1$, $x^t P^m = x^t$. This, coupled with the fact that $x^t P^m = x^t$ has a unique solution (up to a multiplicative constant), implies that all stated in Section 12.1 holds verbatim for P^m and hence for the limit probabilities. This is known as the *irreducible and a-periodic* case. Note, however, one can solve directly for $x^t = x^t P$: the reference for P^m is done for the sake of arguing for this point. An example is given next.

Example. At times, when the stochastic matrix comes with a special structure, it is possible to derive a closed form for its limit probabilities. Consider the following example. Let p_i be with $0 < p_i < 1$, $1 \le i \le n-1$. These numbers can be looked at as probabilities. Note that it is not assumed that they sum up to one. Let the stochastic matrix $P \in R^{n \times n}$ be with $P_{i,i+1} = p_i$ and $P_{i,1} = 1 - p_i$, $1 \le i \le n-1$, and $P_{n1} = 1$. Otherwise, all entries of P equal zero. One can imagine a process which initiates at state 1 and where from state i it moves on to state $i+1$ with probability p_i or returns to the origin with the complementary probability of $1 - p_i$, $1 \le i \le n-1$. Also, state n is the final one in the sense that returning to the origin from there in a single hop is guaranteed. Below see P for the case where $n = 5$

$$P = \begin{pmatrix} 1 - p_1 & p_1 & 0 & 0 & 0 \\ 1 - p_2 & 0 & p_2 & 0 & 0 \\ 1 - p_3 & 0 & 0 & p_3 & 0 \\ 1 - p_4 & 0 & 0 & 0 & p_4 \\ 1 & 0 & 0 & 0 & 0 \end{pmatrix}.$$

Note that P is irreducible and a-periodic. The balance equations are as follows:

$$\pi_1 = \sum_{i=1}^{n-1} \pi_i(1 - p_i) + \pi_n$$

and

$$\pi_i = \pi_{i-1} p_{i-1}, \quad 2 \le i \le n.$$

Solving for this set, one can argue by an induction argument, that the last $n - 1$ equations, lead to

$$\pi_i = \pi_1 \prod_{j=1}^{i-1} p_j, \quad 1 \le i \le n.$$

Using the condition that $\sum_{i=1}^{n} \pi_i = 1$, implies that $\pi_1 = [\sum_{i=1}^{n} \prod_{j=1}^{i-1} p_j]^{-1}$. This completes the solution. Note that one of the balance equations, the first one, is redundant toward this goal. In light of Lemma 12.2 this is not a surprise.

12.4.2 *The unichain and a-periodic case*

What we have here is a single subset of states where a single square sub-matrix associated with the transition probabilities in this subset is irreducible. It is possible to argue that in the case where some of the states do not belong to this subset, then from each one of them there exists a path to one (and hence to all) states in the former subset. The states in this subset are said to be *recurrent*, while all those in the complementary subset are said to be *transient*. The reason behind this terminology is that if i is transient, then sooner or later, it will not be visited again: the process will eventually enter the recurrent subset and will stay there for good. In fact, if T is the square sub-matrix of P representing the transition probabilities among transient states, then $\lim_{m \to \infty} T^m = \underline{0}$, namely T is a transient matrix. Then, possibly after renumbering the states, P comes with the following shape:

$$P = \begin{pmatrix} T & Q \\ \underline{0} & P_1 \end{pmatrix},$$

where P_1 is an irreducible stochastic matrix. Note that

$$P^m = \begin{pmatrix} T^m & Q_m \\ \underline{0} & P_1^m \end{pmatrix},$$

for some non-negative matrix Q_m with $\lim_{m \to \infty} Q_m$ being a non-negative matrix with row-sums being equal to 1. In fact, initializing with $Q_1 = Q$, it is possible to see that $Q_{m+1} = TQ_m + QP_1^m$, $m \geq 1$. Finally, observe that as opposed to the set of recurrent states, the set of transient states might be empty (bringing us back to the irreducible case).

The following lemma will be needed later.

Lemma

Lemma 12.5. *If λ is an eigenvalue of a transient matrix T then $|\lambda| < 1$.*

Proof. Let $v \neq \underline{0}$ be a corresponding right eigenvector, namely $Tv = \lambda v$. Then, $T^m v = \lambda^m v$ for $m \geq 1$. When m goes to infinity the left-hand side goes to $\underline{0}$. Hence, the same is the limit of the right-hand side. Since $v \neq \underline{0}$ this is possible if and only if $|\lambda| < 1$. $\qquad\square$

All said in Section 12.1 holds here too with one exception: $\pi_i = 0$ for all transient states. In particular, if additionally P_1 is irreducible, then there exists a stationary distribution for the recurrent class, denoted by π_1, which obeys $\pi_1^t = \pi_1^t P_1$. Appending this with zeros as the number of transient states prescribes, we will get the limit probabilities of the entire state space.

12.4.3 *The general a-periodic case*

In the general case it is possible that there exists a number, say $m \geq 2$, of subsets, called *equivalence classes*, such that if the process is "trapped" in one of them, it never leaves it. By its definition, all those in the same class, called *recurrent* states are connected to each other in a sense that for any ordered pair of (i, j) among them there exists a path from i to j. For simplicity, assume that each individual class is a-periodic. Also, assume that there exists a (possibly empty) set of transient states where from each one of them there exists a directed path to at least one recurrent state (and hence to all states in the communicating class of the latter). Yet, there is no path from any recurrent state to any of the transient states. The matrix P, maybe after re-numbering the states, has the following shape (for the case where $m = 3$),

$$P = \begin{pmatrix} T & Q_1 & Q_2 & Q_3 \\ \underline{0} & P_1 & \underline{0} & \underline{0} \\ \underline{0} & \underline{0} & P_2 & \underline{0} \\ \underline{0} & \underline{0} & \underline{0} & P_3 \end{pmatrix}.$$

Note that T here is a transient matrix, while P_i are irreducible and a-periodic stochastic matrices, $1 \leq i \leq 3$. Figure 12.1 depicts an example where there are two communicating classes, $\{2, 4\}$ and $\{4, 5\}$, and one transient state, $\{1\}$.

Theorem

Theorem 12.8.

1. *The algebraic and geometric multiplicity of eigenvalue 1 of P coincide. Moreover, they equal m, the number of recurrent classes.*
2. *A basis for the left eigenspace of the eigenvalue 1, is formed of m probability vectors, each of which is full of zeros except for the corresponding entries of the relevant class, where it comes with the limit probabilities of this class when looked at as an isolated process. The same can be said about the right eigenvectors, but the non-zero entries equal one.*

Proof. From Lemma 9.1 we learn that, $P_P(t)$, the characteristic polynomial of P is with

$$P_P(t) = P_T(t) \prod_{i=1}^{m} P_{P_i}(t),$$

from Lemma 12.5 that 1 is not an eigenvalue of T and from Theorem 12.4 we learn that 1 is a simple eigenvalue of P_i, $1 \leq i \leq m$. The fact that m is the algebraic multiplicity of P is now clear. The second part of the Theorem is easily proved by inspection, making the geometric multiplicity of eigenvalue 1 being at least m. By Theorem 9.4 this multiplicity is bounded by the just proved algebraic multiplicity of m. This concludes the proof. □

12.4.4 The periodic and irreducible case

Recall the connectivity assumption made in Section 12.4.1. It is kept here but the a-priodicity assumption is removed. Thus assume that the period equals $d \geq 2$. For the ease of exposition assume that $d = 3$. The general case will then be easily deduced. Then, the state space can be partitioned into three classes, say C_1, C_2 and C_3 such that the process always moves from class C_i to class $C_{(i+1)mod(3)}$. In other words, possibly after re-numbering the states, the matrix P has the following shape:

$$P = \begin{pmatrix} \underline{0} & P_1 & \underline{0} \\ \underline{0} & \underline{0} & P_2 \\ P_3 & \underline{0} & \underline{0} \end{pmatrix},$$

where P_i, $1 \leq i \leq 3$, are not necessarily square matrices, all their entries are non-negative and all their rows sum up to 1. It is then easy to see that

$$P^2 = \begin{pmatrix} \underline{0} & \underline{0} & P_1 P_2 \\ P_2 P_3 & \underline{0} & \underline{0} \\ \underline{0} & P_3 P_1 & \underline{0} \end{pmatrix} \quad \text{and} \quad P^3 = \begin{pmatrix} P_1 P_2 P_3 & \underline{0} & \underline{0} \\ \underline{0} & P_2 P_3 P_1 & \underline{0} \\ \underline{0} & \underline{0} & P_3 P_1 P_2 \end{pmatrix},$$

namely P^3 comes with three equivalence classes. For proof, see, e.g., [1, p. 227]. Then all said in the previous section holds for P^3. Also, note that the stochastic matrix

$$\frac{1}{3}(P + P^2 + P^3)$$

is irreducible and a-periodic, so the results of Section 12.4.1 hold for this (average) matrix.

Chapter 13

Solutions to Exercises

13.1   Chapter 1

Exercise 1.1

Define the vectors: $u = (2, -7, 1)$, $v = (-3, 0, 4)$, $w = (0, 5, -8)$.
Calculate: (a) $3u - 4v$ (b) $2u + 3v - 5w$.

Solution. (a) $3u - 4v = 3 \times (2, -7, 1) - 4 \times (-3, 0, 4) = (6, -21, 3) + (12, 0, -16) = (18, -21, -13)$.   (b) $2u + 3v - 5w = 2 \times (2, -7, 1) + 3 \times (-3, 0, 4) - 5 \times (0, 5, -8) = (4, -14, 2) + (-9, 0, 12) + (0, -25, 40) = (-5, -39, 54)$.

The next two exercises appear in Schaum's series book.

Exercise 1.2

Find x and y that satisfy:
(a) $(x, 3) = (2, x + y)$ (b) $(4, y) = x(2, 3)$.

Solution. (a) From vector equivalence, each element in the left-hand side equals to the corresponding element in the right-hand side. Thus, we obtain the following system of linear equations:

$$x = 2,$$
$$3 = x + y,$$

whose solution is $x = 2$, $y = 1$.

(b) Similar to (a), we obtain the linear system of equations:

$$4 = 2x,$$
$$y = 3x,$$

whose solution is $x = 2, \quad y = 6$.

Exercise 1.3

Normalize the following vectors: (a) $u = (-3, 4)$ (b) $v = (4, -2, -3, 8)$
(c) $w = (\frac{1}{2}, \frac{2}{3}, -\frac{1}{4})$

Solution. We normalize by dividing each of the above vector with its norm.

(a) $\|u\| = \sqrt{u_1^2 + u_2^2} = \sqrt{9 + 16} = 5$. Therefore, the normalized vector is:
$\hat{u} = \frac{1}{5}(-3, 4) = (-\frac{3}{5}, \frac{4}{5})$.
(b) $\|v\| = \sqrt{16 + 4 + 9 + 64} = \sqrt{93} \implies \hat{v} = (\frac{4}{\sqrt{93}}, -\frac{2}{\sqrt{93}}, -\frac{3}{\sqrt{93}}, \frac{8}{\sqrt{93}})$.
(c) Note that normalization of w is invariant to multiple of w with any positive scalar. We can therefore multiply w with 12 to avoid use of fractions. Setting $w' = 12w = (6, 8, -3)$, we have: $\|w'\| = \sqrt{36 + 64 + 9} = \sqrt{109}$. Thus, $\hat{w} = \hat{w}' = (\frac{6}{\sqrt{109}}, \frac{8}{\sqrt{109}}, -\frac{3}{\sqrt{109}})$.

Exercise 1.4

For every $u, v, w \in R^n$ and $k \in R$ prove:
(a) $(u + v)^t w = u^t w + v^t w$ (b) $(ku)^t v = k(u^t v)$ (c) $u^t v = v^t u$

Solution. (a) Denote by u_i, v_i, w_i the ith entry of u, v and w, respectively. It therefore holds that:

$$(u + v)^t w = \sum_{i=1}^{n}(u_i + v_i)w_i = \sum_{i=1}^{n} u_i w_i + v_i w_i$$

$$= \sum_{i=1}^{n} u_i w_i + \sum_{i=1}^{n} v_i w_i = u^t w + v^t w.$$

(b) It holds that:

$$(ku)^t v = \sum_{i=1}^{n}(ku)_i v_i = k \sum_{i=1}^{n} u_i v_i = k(u^t v).$$

(c)

$$u^t v = \sum_{i=1}^{n} u_i v_i = \sum_{i=1}^{n} v_i u_i = v^t u.$$

Exercise 1.5

Find k such that the following vectors are orthogonal: $u = (1, k, -3)$, $v = (2, -5, 4)$.

Solution. Recall that orthogonal vectors $u, v \in R^n$ satisfy $u^t v = 0$. In order to find k, we derive the term for $u^t v$ and equate to zero:

$$u^t v = 2 - 5k - 12 = 0 \implies k = -2.$$

Exercise 1.6

We are given the vectors $a = (2, 5.1, 7)$, $b = (-3, 6.2, 4)$.

(a) Calculate $\|a - \beta b\|^2$ for $\beta = 2$ and $\beta = 0.5$.
(b) Find β which minimizes the above norm.
(c) Find the projection of a on b and its residual.

Solution. (a) For $\beta = 2$ the term $\|a - \beta b\|^2$ becomes:

$$\left\| \begin{pmatrix} 2 \\ 5.1 \\ 7 \end{pmatrix} - 2 \begin{pmatrix} -3 \\ 6.2 \\ 4 \end{pmatrix} \right\|^2 = \left\| \begin{pmatrix} 8 \\ -7.3 \\ -1 \end{pmatrix} \right\|^2 = 118.29,$$

while for $\beta = 0.5$:

$$\left\| \begin{pmatrix} 2 \\ 5.1 \\ 7 \end{pmatrix} - 0.5 \begin{pmatrix} -3 \\ 6.2 \\ 4 \end{pmatrix} \right\|^2 = \left\| \begin{pmatrix} 3.5 \\ 2 \\ 5 \end{pmatrix} \right\|^2 = 41.25.$$

(b) We first derive an equivalent quadratic term for $\|a - \beta b\|^2$:

$$\|a - \beta b\|^2 = (a - \beta b)^t (a - \beta b)$$
$$= a^t a - \beta b^t a - \beta a^t b + \beta^2 b^t b = \beta^2 b^t b - 2\beta b^t a + a^t a.$$

In order to find its minimum, we differentiate w.r.t β and equate to zero:

$$\frac{d}{d\beta} \beta^2 b^t b - 2b^t a + a^t a$$

$$= 2\beta b^t b - 2\beta b^t a = 0 \implies \beta b^t b = b^t a \implies \beta = \frac{b^t a}{b^t b} = \frac{b^t a}{\|b\|^2}.$$

We now plug in a, b to obtain:

$$\beta = \frac{(-3, 6.2, 4)\begin{pmatrix} 2 \\ 5.1 \\ 7 \end{pmatrix}}{(-3)^2 + 6.2^2 + 4^2} = 0.8452.$$

(c) Recall that the projection of a on b is $\frac{b^t a}{\|b\|^2} b$. In our case:

$$0.8452 \begin{pmatrix} -3 \\ 6.2 \\ 4 \end{pmatrix} = \begin{pmatrix} -2.5356 \\ 5.24029 \\ 3.38083 \end{pmatrix}.$$

The residual is

$$a - \frac{b^t a}{\|b\|^2} b = \begin{pmatrix} 2 \\ 5.1 \\ 7 \end{pmatrix} - \begin{pmatrix} -2.5356 \\ 5.24029 \\ 3.38083 \end{pmatrix} = \begin{pmatrix} 4.53562421 \\ -0.14029004 \\ 3.61916772 \end{pmatrix}.$$

13.2   Chapter 2

Exercise 2.1

Prove that all vectors which are orthogonal to a vector v form a linear subspace.

Solution. Recall that if u, w belong to some linear subspace, then for every $\alpha, \beta \in R$ also $\alpha u + \beta w$ belongs to that linear subspace. In our case we would like to show that if u and w are orthogonal to v then also $\alpha u + \beta w$ is orthogonal to v, for every $\alpha, \beta \in R$. Indeed, $(\alpha u + \beta w)^t v = \alpha u^t v + \underbrace{\beta w^t v}_{(i)} = \alpha \times 0 + \beta \times 0 = 0$, where equality (i) follows from the assumption that u, w are orthogonal to v.

Exercise 2.2

Let V_1, V_2 be linear subspaces.

(a) Prove that the intersection of V_1 and V_2 form a linear subspace.
(b) Show by a counter-example that the union of V_1 and V_2 does not necessarily form a linear subspace.

Solution. (a) Define $W = V_1 \cap V_2$. Since each of V_1 and V_2 is a linear subspace, then $0 \in V_1$ and $0 \in V_2$, which implies that $0 \in V_1 \cap V_2 = W$, and so W is not an empty set. Next, suppose that $w \in W$. This implies that $w \in V_1$ and $w \in V_2$, which in turn implies that $\alpha w \in V_1$ and $\alpha w \in V_2$ for every $\alpha \in R$, from linear subspace properties. Thus, $\alpha w \in W$ for every $\alpha \in R$. Last, suppose that $w_1, w_2 \in W$. Then $w_1, w_2 \in V_1$ and $w_1, w_2 \in V_2$. Since each of V_1 and V_2 is a linear subspace, then $w_1 + w_2 \in V_1$ and $w_1 + w_2 \in V_2$, and therefore $w_1 + w_2 \in W$.

(b) Consider $V_1 = \{(\alpha, 0) | \alpha \in R\}$ and $V_2 = \{(0, \alpha) | \alpha \in R\}$. It is easy to show that each set is a linear subspace (note that V_1 consists of all linear combinations of the vector $(1, 0)$. Likewise, V_2 consists of all linear combinations of the vector $(0, 1)$. See example 2 in Section 2.2. Then clearly $(1, 0) \in V_1$ and $(0, 1) \in V_2$ but their sum, $(1, 1)$ is not in $V_1 \cup V_2$. We found an example where $v_1 \in V_1$ and $v_2 \in V_2$, so that $v_1, v_2 \in V_1 \cup V_2$ but $v_1 + v_2 \notin V_1 \cup V_2$, and therefore $V_1 \cup V_2$ is not a linear subspace.

Exercise 2.3

We are given n vectors $x_1, \ldots, x_n \in R^n$ which are all non-zero, such that $x_1, \ldots, x_{n-1}$ are linearly independent. Prove that the vectors $x_1, \ldots, x_n$ are linearly independent if and only if x_n cannot be written as a linear combination of $x_1, \ldots, x_{n-1}$.

Solution. By way of contradiction, we will prove the equivalent statement that $x_1, \ldots, x_n$ are not linearly independent if and only if x_n can be written as a linear combination of $x_1, \ldots, x_{n-1}$. First, suppose that $x_1, \ldots, x_n$ are not linearly independent. Then there exist $\alpha_1, \ldots, \alpha_n \in R$, not all of them being zero, such that $\sum_{i=1}^{n} \alpha_i x_i = 0$. Now, since it is assumed that $x_1, \ldots, x_{n-1}$ are linearly independent, then α_n must be non-zero (otherwise it holds that $\sum_{i=1}^{n-1} \alpha_i x_i = 0$, where not all $\alpha_1, \ldots, \alpha_{n-1}$ are zero, which contradicts the above assumption). We can therefore isolate x_n and write it as: $x_n = -\sum_{i=1}^{n-1} \frac{\alpha_i}{\alpha_n} x_i$, so that x_n can be written as a linear combination of $x_1, \ldots, x_{n-1}$. We now prove the reverse direction, which is trivial; if x_n can be written as a linear combination: $x_n = \sum_{i=1}^{n-1} \alpha_i x_i$, then clearly $x_n - \sum_{i=1}^{n-1} \alpha_i x_i = 0$. Note that the first coefficient in the latter linear combination is 1. Thus, $x_1, \ldots, x_n$ are not linearly independent.

Exercise 2.4

Let V, W be linear subspaces of R^n.

(a) Define $V + W = \{v + w | v \in V, w \in W\}$. Prove that $V + W$ is a linear subspace.

(b) Prove that if V and W don't have any common element except from the zero vector, then it holds that $\dim(V + W) = \dim(V) + \dim(W)$.

Solution. (a) Since $0 \in V$ and $0 \in W$ then $0 = 0 + 0 \in V + W$ and therefore $V + W$ is non-empty. Now, for any $\alpha \in R$ note that $\alpha(v + w) = \alpha v + \alpha w \in V + W$, since $\alpha v \in V$ and $\alpha w \in W$. Finally, taking $v_i + w_i$ for $i = 1, 2$, it holds that $(v_1 + w_1) + (v_2 + w_2) = (v_1 + v_2) + (w_1 + w_2) \in V + W$, since $\forall v_1, v_2 \in V \implies v_1 + v_2 \in V$, as well as $\forall w_1, w_2 \in W \implies w_1 + w_2 \in W$.

(b) Let $\{v_1, \ldots, v_k\}$ be the basis of V and $\{w_1, \ldots, w_\ell\}$ be the basis of W. That is, $\dim(V) = k$ and $\dim(W) = \ell$. We then can write any vector $u \in V + W$ as: $u = \sum_{i=1}^{k} \alpha_i v_i + \sum_{j=1}^{\ell} \beta_j w_j$, where $\alpha_i, \beta_j \in R$ for all $i = 1, \ldots, k$, $j = 1, \ldots, \ell$. In other words, the set $\{v_1, \ldots, v_k, w_1, \ldots, w_\ell\}$ spans $V + W$. Next we show that all these vectors are linearly independent. By contradiction, if these vectors are not linearly independent, then there exists a non-trivial solution to $\sum_{i=1}^{k} \alpha_i v_i + \sum_{j=1}^{\ell} \beta_j w_j = 0$, which can also be written as $\sum_{i=1}^{k} \alpha_i v_i = -\sum_{j=1}^{\ell} \beta_j w_j$. The vector in the left-hand side is in V, while the one in the right-hand side is in W. Since they are identical, both are in $V \cap W$. But we know that the intersection consists only of the zero vector, and therefore $\sum_{i=1}^{k} \alpha_i v_i = 0$. Because $v_1, \ldots, v_k$ are linearly independent (recall that they form the basis of V), then $\alpha_1 = \cdots = \alpha_k = 0$. Likewise, $-\sum_{j=1}^{\ell} \beta_j w_j$, and since $w_1, \ldots, w_\ell$ are linearly independent, then $\beta_1 = \cdots = \beta_\ell = 0$. We found that if $\sum_{i=1}^{k} \alpha_i v_i + \sum_{j=1}^{\ell} \beta_j w_j = 0$ then $\alpha_1 = \cdots = \alpha_k = \beta_1 = \cdots = \beta_\ell = 0$, and therefore the vectors in the set $\{\alpha_1, \ldots, \alpha_k, \beta_1, \ldots, \beta_\ell\}$ are linearly independent. Thus, $\{\alpha_1, \ldots, \alpha_k, \beta_1, \ldots, \beta_\ell\}$ is the linear basis of $V + W$, and its dimension is $\dim(V + W) = k + \ell$.

Exercise 2.5

Prove that the set of vectors $\{f_i\}_{i=1}^{n}$, such that $f_i \in R^n$ and $f_i = \Sigma_{j=1}^{i} e_j$, $1 \leq i \leq n$, forms a linear basis of R^n.

Solution. For any vector $v \in R^n$, our goal is to show that it can be written as a linear combination $\sum_{i=1}^{n} \alpha_i f_i$, for $\alpha_i \in R$, $\forall i = 1, \ldots, n$. Note that $f_1 = e_1$, and $e_i = f_i - f_{i-1}$ for $i \geq 2$. Thus, we have: $v = \sum_{i=1}^{n} v_i e_i = v_1 f_1 + \sum_{i=2}^{n} v_i (f_i - f_{i-1}) = \sum_{i=1}^{n-1} (v_i - v_{i+1}) f_i + v_n f_n$. We showed that any vector $v \in R^n$ can be written as a linear combination of $\{f_i\}_{i=1}^{n}$. Thus, R^n is spanned by $\{f_i\}_{i=1}^{n}$, whose dimension is n.

13.3 Chapter 3

Exercise 3.1

(a) Show that the set of all vectors which are orthogonal to any vector in a linear subspace $V \subset R^n$, is a linear subspace itself. Denote it by V^+.

(b) Show that $(V^+)^+ = V$ (*hint*: use the Gram–Schmidt process).

(c) Show that for any given vector v in R^n, the residual of its projection on V, is the projection on V^+.

Solution. (a) The zero vector is orthogonal to all vectors in V, and therefore $0 \in V^+$, thus V^+ is non-empty. Now, if $v_1, v_2 \in V^+$, then $v_1^t u = 0$ and $v_2^t u = 0$ for any $u \in V$. It is clear that for any scalar $\alpha, \beta \in R$ we have $(\alpha v_1 + \beta v_2)^t u = \alpha v_1^t u + \beta v_2^t u = 0 + 0 = 0$, and therefore $\alpha v_1 + \beta v_2 \in V^+$. This completes the proof.

(b) Since $V \subset R^n$ is a linear subspace, we can construct an orthonormal basis using the Gram–Schmidt process. Denote that orthonormal basis by $V = span\{v_1, \ldots, v_k\}$ (assuming that $dim(V) = k$). Let us construct now an orthonormal basis for R^n that includes $v_1, \ldots, v_k$. For instance, we can begin with the set of vectors $\{v_1, \ldots, v_k, e_1, \ldots, e_n\}$ which clearly spans R^n. Running the Gram–Schmidt process at that order, the vectors $v_1, \ldots, v_k$ stay intact, since they are orthonormal. Thus, we end up with an orthonormal basis $\{v_1, \ldots, v_k, v_{k+1}, \ldots, v_n\}$. We will show now that $V^+ = span\{v_{k+1}, \ldots, v_n\}$, by proving that (i) $span\{v_{k+1}, \ldots, v_n\} \subseteq V^+$ and (ii) $V^+ \subseteq span\{v_{k+1}, \ldots, v_n\}$. The first direction (i) is trivial; if $u \in span\{v_{k+1}, \ldots, v_n\}$ then it can be written as a linear combination $\sum_{i=k+1}^{n} \alpha_i v_i$, and it is orthogonal to any linear combination $\sum_{i=1}^{k} \alpha_i v_i$, i.e., any vector in $span\{v_1, \ldots, v_k\}$. We prove the second direction (ii) through negation. If $u \notin span\{v_{k+1}, \ldots, v_n\}$ then it must include some/all of $v_1, \ldots, v_k$ in its representation as a linear combination; i.e., $u = \sum_{i=1}^{n} \alpha_i v_i$ where $\alpha_i \neq 0$ for at least one $1 \leq i \leq k$. Assume, without loss of generality, that $\alpha_1 \neq 0$. Then $u^t v_1 = \alpha_1 v_1^t v_1 = \alpha_1 \neq 0$ (recall that $v_1, \ldots, v_n$ are all orthonormal). Thus, u is not orthogonal to all vectors in V, and $u \notin V^+$. We now complete the proof by showing that $(V^+)^+ = span\{v_1, \ldots, v_k\}$

(recall that $V = span\{v_1, \ldots, v_k\}$), proving again two directions. (i) If $u \in span\{v_1, \ldots, v_k\}$ then $u = \sum_{i=1}^{k} \alpha_i v_i$ which is orthogonal to any vector in $span\{v_{k+1}, \ldots, v_n\} = V^+$, proving that $span\{v_1, \ldots, v_k\} \subseteq (V^+)^+$. (ii) Through negation, if $u \notin span\{v_1, \ldots, v_k\}$, then as a linear combination of the orthonormal basis $u = \sum_{i=1}^{n} \alpha_i v_i$ it includes at least one non-zero coefficient $\alpha_{k+1}, \ldots, \alpha_n$. Assume without loss of generality that $\alpha_{k+1} \neq 0$, then $u^t v_{k+1} = \alpha_{k+1} \neq 0$, and it is not orthogonal to all vectors in V^+. Thus, $u \notin (V^+)^+$. From (i)–(ii) we conclude that $(V^+)^+ = V$.

(c) Using the notations in (b), the vector v can be written as $v = \sum_{i=1}^{n} \alpha_i v_i$. Note that since $\{v_1, \ldots, v_n\}$ is an orthonormal basis of R^n, then $\alpha_i = v^t v_i$ (Theorem 3.2.). Moreover, from Theorem 3.3. it follows that the projection of v on $V = span\{v_1, \ldots, v_k\}$ equals to $\sum_{i=1}^{k} \alpha_i v_i$. Thus, the residual of its projection on V is $v - \sum_{i=1}^{k} \alpha_i v_i = \sum_{i=k+1}^{n} \alpha_i v_i$, which is the projection of v on V^+ (Theorem 3.3). Recall that $V^+ = span\{v_{k+1}, \ldots, v_n\}$.

Exercise 3.2

Consider R^4 and the subspace $V = span\{v_1, v_2, v_3\}$, where $v_1 = (1, 1, 1, 1)$, $v_2 = (1, 1, 2, 4)$ and $v_3 = (1, 2, -4, -3)$. Find an orthonormal basis for V using the Gram–Schmidt process.

Solution. Through the Gram–Schmidt process we will find an orthonormal basis $\{w_1, w_2, w_3\}$. We first set:

$$w_1 = \frac{v_1}{\|v_1\|} = \frac{1}{\sqrt{1+1+1+1}} \begin{pmatrix} 1 \\ 1 \\ 1 \\ 1 \end{pmatrix} = \begin{pmatrix} 1/2 \\ 1/2 \\ 1/2 \\ 1/2 \end{pmatrix}.$$

Next,

$$u_2 = v_2 - (v_2^t w_1) w_1$$

$$= \begin{pmatrix} 1 \\ 1 \\ 2 \\ 4 \end{pmatrix} - \left(\begin{pmatrix} 1 & 1 & 2 & 4 \end{pmatrix} \begin{pmatrix} 1/2 \\ 1/2 \\ 1/2 \\ 1/2 \end{pmatrix} \right) \begin{pmatrix} 1/2 \\ 1/2 \\ 1/2 \\ 1/2 \end{pmatrix} = \begin{pmatrix} 1 \\ 1 \\ 2 \\ 4 \end{pmatrix} - \begin{pmatrix} 2 \\ 2 \\ 2 \\ 2 \end{pmatrix} = \begin{pmatrix} -1 \\ -1 \\ 0 \\ 2 \end{pmatrix},$$

so we get

$$w_2 = \frac{u_2}{\|u_2\|} = \frac{1}{\sqrt{1+1+0+4}} \begin{pmatrix} -1 \\ -1 \\ 0 \\ 2 \end{pmatrix} = \frac{1}{\sqrt{6}} \begin{pmatrix} -1 \\ -1 \\ 0 \\ 2 \end{pmatrix}.$$

Finally, $u_3 = v_3 - (v_3^t w_1)w_1 - (v_3^t w_2)w_2$. Plugging in the vectors w_1, w_2 and v_3, we arrive at:

$$u_3 = \begin{pmatrix} 1 \\ 2 \\ -4 \\ -3 \end{pmatrix} - (-2) \begin{pmatrix} 1/2 \\ 1/2 \\ 1/2 \\ 1/2 \end{pmatrix} - \left(-\frac{9}{\sqrt{6}} \right) \cdot \frac{1}{\sqrt{6}} \begin{pmatrix} -1 \\ -1 \\ 0 \\ 2 \end{pmatrix} = \begin{pmatrix} 1/2 \\ 3/2 \\ -3 \\ 1 \end{pmatrix}.$$

Thus,

$$w_3 = \frac{1}{\sqrt{\frac{1}{4} + \frac{9}{4} + 9 + 1}} \begin{pmatrix} 1/2 \\ 3/2 \\ -3 \\ 1 \end{pmatrix} = \frac{1}{\sqrt{12.5}} \begin{pmatrix} 1/2 \\ 3/2 \\ -3 \\ 1 \end{pmatrix}.$$

13.4   Chapter 4

Exercise 4.1

Let $V = U = R^3$. Check whether or not $f : U \to V$ is a linear function, where:

$$T \begin{pmatrix} x \\ y \\ z \end{pmatrix} = \begin{pmatrix} x+y \\ 0 \\ 2x+z \end{pmatrix}.$$

Solution. In order to find whether or not T is linear we need to check if the following holds: $T(\alpha a + \beta b) = \alpha T(a) + \beta T(b)$. Indeed,

$$T(\alpha a + \beta b) = T \left(\begin{pmatrix} \alpha x_1 \\ \alpha y_1 \\ \alpha z_1 \end{pmatrix} + \begin{pmatrix} \beta x_2 \\ \beta y_2 \\ \beta z_2 \end{pmatrix} \right)$$

$$= T \left(\begin{pmatrix} \alpha x_1 + \beta x_2 \\ \alpha y_1 + \beta y_2 \\ \alpha z_1 + \beta z_2 \end{pmatrix} \right) = \begin{pmatrix} \alpha x_1 + \beta x_2 + \alpha y_1 + \beta y_2 \\ 0 \\ 2(\alpha x_1 + \beta x_2) + \alpha z_1 + \beta z_2 \end{pmatrix}$$

$$= \alpha \begin{pmatrix} x_1 + y_1 \\ 0 \\ 2x_1 + z_1 \end{pmatrix} + \beta \begin{pmatrix} x_2 + y_2 \\ 0 \\ 2x_2 + z_2 \end{pmatrix} = \alpha T(a) + \beta T(b).$$

Exercise 4.2

Let $V = W = R^2$. Prove or disprove: the function f, which is defined as follows, is linear:

$$f \begin{pmatrix} a \\ b \end{pmatrix} = \begin{pmatrix} a^2 \\ b \end{pmatrix}.$$

Solution. The claim is not true. We will show that by a counter-example. If f is a linear function, then it must satisfy $f(\alpha v) = \alpha f(v)$. However, if

$a = b = 1$ and $\alpha = 3$ we get:

$$f\left(3\left(\begin{smallmatrix}1\\1\end{smallmatrix}\right)\right) = f\left(\begin{smallmatrix}3\\3\end{smallmatrix}\right) = \left(\begin{smallmatrix}9\\3\end{smallmatrix}\right) \neq \left(\begin{smallmatrix}3\\3\end{smallmatrix}\right) = 3f\left(\begin{smallmatrix}1\\1\end{smallmatrix}\right).$$

13.5 Chapter 5

Exercise 5.1

Let $A \in R^{m \times n}$ and $f : R^n \rightarrow R^m$, such that the vector $f(x)$ is defined as $(f(x))_i = \sum_{j=1}^{n} A_{ij} x_j$. Prove that $f(x)$ is a linear function.

Solution. In order to check whether or not f is a linear function, let us first find out if $(f(\alpha x + \beta y))_i = \alpha(f(x))_i + \beta(f(y))_i$. Indeed,

$$(f(\alpha x + \beta y))_i = \sum_{j=1}^{n} A_{ij}(\alpha x + \beta y)_j = \sum_{j=1}^{n} A_{ij}(\alpha x_j + \beta y_j)$$

$$= \alpha \sum_{j=1}^{n} A_{ij} x_j + \beta \sum_{j=1}^{n} B_{ij} y_j = \alpha(f(x))_i + \beta(f(y))_j.$$

Exercise 5.2

The following function is linear (see Exercise 4.1):

$$T\left(\begin{smallmatrix}x\\y\\z\end{smallmatrix}\right) = \left(\begin{smallmatrix}x+y\\0\\2x+z\end{smallmatrix}\right).$$

Find the matrix A which represents the above function. For that purpose, calculate $T(e_i)$ for $1 \leq i \leq 3$, and calculate $\sum_{i=1}^{3} x_i T(e_i)$.

Solution. Since T is linear, we have:

$$T\left(\begin{smallmatrix}x\\y\\z\end{smallmatrix}\right) = xT(e_1) + yT(e_2) + zT(e_3) = \begin{pmatrix} \vdots & \vdots & \vdots \\ T(e_1) & T(e_2) & T(e_3) \\ \vdots & \vdots & \vdots \end{pmatrix} \left(\begin{smallmatrix}x\\y\\z\end{smallmatrix}\right).$$

We will therefore obtain the matrix A by calculating:

$$T(e_1) = T\left(\begin{smallmatrix}1\\0\\0\end{smallmatrix}\right) = \left(\begin{smallmatrix}1\\0\\2\end{smallmatrix}\right),$$

$$T(e_2) = T\left(\begin{smallmatrix}0\\1\\0\end{smallmatrix}\right) = \left(\begin{smallmatrix}1\\0\\0\end{smallmatrix}\right),$$

$$T(e_3) = T\left(\begin{smallmatrix}0\\0\\1\end{smallmatrix}\right) = \left(\begin{smallmatrix}0\\0\\1\end{smallmatrix}\right),$$

and therefore,

$$A = \begin{pmatrix} 1 & 1 & 0 \\ 0 & 0 & 0 \\ 2 & 0 & 1 \end{pmatrix}.$$

Exercise 5.3

(a) Prove that the residual of the projection of x on b is a linear function.
(b) Find the representative matrix A of the linear function specified in (a).

Solution. (a) As we know, the residual of the projection of x on b is given by $f(x) = x - \frac{x^t b}{\|b\|^2} b$. We will show that $f(\alpha x_1 + \beta x_2) = \alpha f(x_1) + \beta f(x_2)$:

$$f(\alpha x_1 + \beta x_2) = (\alpha x_1 + \beta x_2) - \frac{(\alpha x_1 + \beta x_2)^t b}{\|b\|^2} b$$

$$= (\alpha x_1 + \beta x_2) - \frac{(\alpha x_1)^t b}{\|b\|^2} b - \frac{(\beta x_2)^t b}{\|b\|^2} b$$

$$= \alpha x_1 - \frac{\alpha x_1^t b}{\|b\|^2} b + \beta x_2 - \frac{\beta x_2^t b}{\|b\|^2} b$$

$$= \alpha \left(x_1 - \frac{x_1^t b}{\|b\|^2} b \right) + \beta \left(x_2 - \frac{x_2^t b}{\|b\|^2} b \right) = \alpha f(x_1) + \beta f(x_2).$$

(b) Notice that $f(e_i) = e_i - \frac{e_i^t b}{\|b\|^2} b = e_i - \frac{b_i}{\|b\|^2} b$, and therefore:

$$f(x) = \begin{pmatrix} \vdots & \vdots & & \vdots \\ f(e_1) & f(e_2) & \cdots & f(e_n) \\ \vdots & \vdots & & \vdots \end{pmatrix} \begin{pmatrix} x_1 \\ x_2 \\ \vdots \\ x_n \end{pmatrix}$$

$$= \begin{pmatrix} 1 - \frac{b_1^t b_1}{\|b\|^2} & 0 - \frac{b_2^t b_1}{\|b\|^2} & \cdots & 0 - \frac{b_n^t b_1}{\|b\|^2} \\ 0 - \frac{b_1^t b_2}{\|b\|^2} & 1 - \frac{b_2^t b_2}{\|b\|^2} & \cdots & 0 - \frac{b_n^t b_2}{\|b\|^2} \\ \vdots & \vdots & \cdots & \vdots \\ 0 - \frac{b_1^t b_n}{\|b\|^2} & 0 - \frac{b_2^t b_n}{\|b\|^2} & \cdots & 1 - \frac{b_n^t b_n}{\|b\|^2} \end{pmatrix} \begin{pmatrix} x_1 \\ x_2 \\ \vdots \\ x_n \end{pmatrix} = \left(I - \frac{bb^t}{\|b\|^2} \right) \begin{pmatrix} x_1 \\ x_2 \\ \vdots \\ x_n \end{pmatrix}.$$

Thus, the matrix A which represents the linear transformation from (a) is $A = I - \frac{bb^t}{\|b\|^2}$.

Exercise 5.4

Suppose that $F(x) : R^n \to R^n$ satisfies $F_i(x) = x_i - \bar{x}$, such that $\bar{x} = \frac{1}{n}\sum_{i=1}^{n} x_i$.

(a) Prove that $F(x)$ is a linear function.
(b) What is the representative matrix A of the function $F(x)$, with respect to the standard basis?
(c) What is $dim(null(F))$?

Solution. (a) We will show that $F(\alpha x + \beta y) = \alpha F(x) + \beta F(y)$, entry-wise:

$$F_i(\alpha x + \beta y) = (\alpha x + \beta y)_i - \overline{\alpha x + \beta y}$$

$$= \alpha x_i + \beta y_i - \frac{1}{n}\sum_{i=1}^{n}(\alpha x + \beta y)_i$$

$$= \alpha x_i + \beta y_i - \frac{1}{n}\sum_{i=1}^{n}\alpha x_i - \frac{1}{n}\sum_{i=1}^{n}\beta y_i$$

$$\alpha\left(x_i - \frac{1}{n}\sum_{i=1}^{n}x_i\right) + \beta\left(y_i - \frac{1}{n}\sum_{i=1}^{n}y_i\right) = \alpha F_i(x) + \beta F_i(y).$$

(b) For the ith member e_i of the standard basis we have $\bar{e_i} = \frac{1}{n}$. Then, it is clear that $F(e_i) = e_i - \frac{1}{n}$. We therefore construct the matrix A as follows:

$$A = \begin{pmatrix} \vdots & & \vdots \\ F(e_1) & \cdots & F(e_n) \\ \vdots & & \vdots \end{pmatrix} = \begin{pmatrix} 1-\frac{1}{n} & \cdots & -\frac{1}{n} \\ \vdots & \ddots & \vdots \\ -\frac{1}{n} & \cdots & 1-\frac{1}{n} \end{pmatrix},$$

i.e., all A's entries are $-\frac{1}{n}$, except the main diagonal, whose entries are $1 - \frac{1}{n}$.

(c) A vector v in $null(F)$ satisfies $v_i - \bar{v} = 0$ for each of $i = 1, \ldots, n$. Since $\bar{v}$ is a constant, this means that $v_1 = \cdots = v_n$. That is, all entries of v must be identical, and the vector v is written as $v = \alpha(1, \ldots, 1)$, where α is a scalar. In fact, it is easy to see the $v \in null(F)$ for any $\alpha \in R$, and therefore $dim(null(F)) = 1$, since $null(F)$ is a linear combination of $(1, \ldots, 1)$.

Exercise 5.5

Let $A = \left(\begin{smallmatrix} 4 & 1 & 0 \\ 5 & 8 & 3 \end{smallmatrix}\right)$ and $B = \left(\begin{smallmatrix} 1 & 9 & 3 & 1 \\ 7 & 2 & 8 & 1 \\ 4 & 0 & 6 & 5 \end{smallmatrix}\right)$. Compute AB.

Solution. In order to obtain the ijth entry of matrix AB, we need to calculate the inner product between the ith row of A and the jth column of B. We therefore get:

$$AB = \begin{pmatrix} 11 & 38 & 20 & 5 \\ 73 & 61 & 97 & 28 \end{pmatrix}.$$

Exercise 5.6

Suppose that $A \in R^{m \times k}$ and $B \in R^{k \times n}$.

(a) For $C \in R^{k \times n}$, prove that $A(B + C) = AB + AC$.
(b) For $C \in R^{n \times p}$, prove that $(AB)C = A(BC)$.

Solution. (a) Following again Definition 5.1, we have $(A(B + C))_{ij} = \sum_{\ell=1}^{k} A_{i\ell}(B+C)_{\ell j} = \sum_{\ell=1}^{k} A_{i\ell}(B_{\ell j}+C_{\ell j}) = \sum_{\ell=1}^{k} A_{i\ell}B_{\ell j}+\sum_{\ell=1}^{k} A_{i\ell}C_{\ell j} = AB + AC$.

(b) It holds that $((AB)C)_{ij} = \sum_{\ell=1}^{k}(AB)_{i\ell}C_{\ell j} = \sum_{\ell=1}^{k}(\sum_{h=1}^{k} A_{ih}B_{h\ell})C_{\ell j}$. The right-hand side can be written as:

$$\sum_{\ell=1}^{k}(A_{i1}B_{1\ell} + \cdots + A_{ik}B_{k\ell})C_{\ell j} = \sum_{\ell=1}^{k} A_{i1}B_{1\ell}C_{\ell j} + \cdots + \sum_{\ell=1}^{k} A_{ik}B_{k\ell}C_{\ell j}$$

$$= A_{i1} \sum_{\ell=1}^{k} B_{1\ell}C_{\ell j} + \cdots + A_{ik} \sum_{\ell=1}^{k} B_{k\ell}C_{\ell j}$$

$$= A_{i1}(BC)_{1j} + \cdots + A_{ik}(BC)_{kj} = \sum_{h=1}^{k} A_{ih}(BC)_{hj} = A(BC).$$

Exercise 5.7

For $A, B \in R^{m \times n}$ and $C \in R^{n \times k}$ prove that:

(a) $(A + B)C = AC + BC$.
(b) $(A^t B)C = A^t(BC) = A^t BC$.

Solution. (a) Indeed, for each i, jth entry we have:

$$((A+B)C)_{ij} = \sum_{\ell=1}^{n}(A+B)_{i\ell}C_{\ell j} = \sum_{\ell=1}^{n}(A_{i\ell}C_{\ell j} + B_{i\ell}C_{\ell j})$$

$$= \sum_{\ell=1}^{n}A_{i\ell}C_{\ell j} + \sum_{\ell=1}^{n}B_{i\ell}C_{\ell j} = (AC)_{ij} + (BC)_{ij}.$$

(b) Again, regarding the i, jth entry, note that $(A^t B)_{ij} = \sum_{\ell=1}^{m} A_{\ell i}B_{\ell j}$ and $A^t B \in R^{n \times n}$. We have:

$$((A^t B)C)_{ij} = \sum_{\ell=1}^{n}(A^t B)_{i\ell}C_{\ell j} = \sum_{\ell=1}^{n}\sum_{h=1}^{m}A_{hi}B_{h\ell}C_{\ell j}.$$

On the other hand, it also holds that

$$(A^t(BC))_{ij} = \sum_{h=1}^{m}A_{hi}(BC)_{hj} = \sum_{h=1}^{m}A_{hi}\sum_{\ell=1}^{n}B_{h\ell}C_{\ell j} = \sum_{\ell=1}^{n}\sum_{h=1}^{m}A_{hi}B_{h\ell}C_{\ell j},$$

and therefore $((A^t B)C)_{ij}$ and $(A^t(BC))_{ij}$ are identical.

Exercise 5.8

Let

$$A = \begin{pmatrix} 2 & 0 & 0 \\ 0 & 4 & 1 \\ 0 & 0 & 9 \\ 1 & 7 & 0 \\ 0 & 0 & -3 \end{pmatrix}.$$

(a) Perform the Gram–Schmidt process on the columns of A.
(b) Find the QR factorization of A.

Solution. (a) Denote by $a_i \in R^5$ the column vectors of A, for $i = 1, 2, 3$. Denote by $u_i \in R^3$ the orthogonal column vectors of A, and by $v_i \in R^3$ the orthonormal column vectors of A, for $i = 1, 2, 3$. We first set $u_1 = a_1$. It is left to compute u_2 and u_3, using the Gram–Schmidt process:

$$u_2 = a_2 - \text{Proj}_{u_1}(a_2) = a_2 - \frac{a_2^t u_1}{\|u_1\|^2} u_1 = \begin{pmatrix} 0 \\ 4 \\ 0 \\ 7 \\ 0 \end{pmatrix} - \frac{(0\ 4\ 0\ 7\ 0)\begin{pmatrix} 2 \\ 0 \\ 0 \\ 1 \\ 0 \end{pmatrix}}{5} \begin{pmatrix} 2 \\ 0 \\ 0 \\ 1 \\ 0 \end{pmatrix}$$

$$= \begin{pmatrix} 0 \\ 4 \\ 0 \\ 7 \\ 0 \end{pmatrix} - \frac{7}{5}\begin{pmatrix} 2 \\ 0 \\ 0 \\ 1 \\ 0 \end{pmatrix} = \begin{pmatrix} -2.8 \\ 4 \\ 0 \\ 5.6 \\ 0 \end{pmatrix},$$

$$u_3 = a_3 - \text{Proj}_{u_1}(a_3) - \text{Proj}_{u_2}(a_3) = a_3 - \frac{a_3^t u_1}{\|u_1\|^2} u_1 - \frac{a_3^t u_2}{\|u_2\|^2} u_2$$

$$= \begin{pmatrix} 0 \\ 1 \\ 9 \\ 0 \\ -3 \end{pmatrix} - \frac{(0\ 1\ 9\ 0\ -3)\begin{pmatrix} 2 \\ 0 \\ 0 \\ 1 \\ 0 \end{pmatrix}}{5} \begin{pmatrix} 2 \\ 0 \\ 0 \\ 1 \\ 0 \end{pmatrix} - \frac{(0\ 1\ 9\ 0\ -3)\begin{pmatrix} -2.8 \\ 4 \\ 0 \\ 5.6 \\ 0 \end{pmatrix}}{55.2} \begin{pmatrix} -2.8 \\ 4 \\ 0 \\ 5.6 \\ 0 \end{pmatrix}$$

$$= \begin{pmatrix} 0 \\ 1 \\ 9 \\ 0 \\ -3 \end{pmatrix} - \frac{4}{55.2}\begin{pmatrix} -2.8 \\ 4 \\ 0 \\ 5.6 \\ 0 \end{pmatrix} = \begin{pmatrix} 0.2028986 \\ 0.7101449 \\ 9 \\ -0.4057971 \\ -3 \end{pmatrix}.$$

Now we can compute the orthonormal vectors v_1, v_2, v_3:

$$v_1 = \frac{u_1}{\|u_1\|} = \frac{1}{\sqrt{5}}\begin{pmatrix} 2 \\ 0 \\ 0 \\ 1 \\ 0 \end{pmatrix} = \begin{pmatrix} 2 \\ 0 \\ 0 \\ 1 \\ 0 \end{pmatrix} = \begin{pmatrix} 0.8944272 \\ 0 \\ 0 \\ 0.4472136 \\ 0 \end{pmatrix},$$

$$v_2 = \frac{u_2}{\|u_2\|} = \begin{pmatrix} -0.3768673 \\ 0.5383819 \\ 0 \\ 0.7537347 \\ 0 \end{pmatrix},$$

$$v_3 = \frac{u_3}{\|u_3\|} = \begin{pmatrix} 0.02130351 \\ 0.07456226 \\ 0.94496251 \\ -0.04260701 \\ -0.31498750 \end{pmatrix}.$$

(b) From (a) we know that Q equals to:

$$Q = \begin{pmatrix} 0.8944272 & -0.3768673 & 0.02130351 \\ 0 & 0.5383819 & 0.07456226 \\ 0 & 0 & 0.94496251 \\ 0.4472136 & 0.7537347 & -0.04260701 \\ 0 & 0 & -0.31498750 \end{pmatrix}.$$

The matrix R is obtained through $R = Q^t A$:

$$R = Q^t A$$

$$= \begin{pmatrix} 0.894427 & 0 & 0 & 0.4472136 & 0 \\ -0.3768673 & 0.5383819 & 0 & 0.7537347 & 0 \\ 0.02130351 & 0.07456226 & 0.94496251 & -0.04260701 & -0.31498750 \end{pmatrix} \begin{pmatrix} 2 & 0 & 0 \\ 0 & 4 & 1 \\ 0 & 0 & 9 \\ 1 & 7 & 0 \\ 0 & 0 & -3 \end{pmatrix},$$

which becomes:

$$R = \begin{pmatrix} 2.2361 & 3.1305 & 0 \\ 0 & 7.4297 & 0.5384 \\ 0 & 0 & 9.5242 \end{pmatrix}.$$

Exercise 5.9

Compute the QR factorization of the following matrix:

$$A = \begin{pmatrix} 12 & -51 & 4 \\ 6 & 167 & -68 \\ -4 & 24 & -41 \end{pmatrix}.$$

Solution. Let us find an orthonormal basis for the columns of A through the Gram–Schmidt process in order to find the matrix Q. After this we find a matrix R via the equation $R = Q^t A$. Denote by $a_i \in R^3$ the column vectors of A, for $i = 1, 2, 3$. Denote by $u_i \in R^3$ the orthogonal column vectors of A, and by $v_i \in R^3$ the orthonormal column vectors of A, for $i = 1, 2, 3$. We begin with the Gram–Schmidt process:

$$u_1 = a_1 = \begin{pmatrix} 12 \\ 6 \\ -4 \end{pmatrix},$$

$$u_2 = a_2 - \text{Proj}_{u_1}(a_2) = a_2 - \frac{a_2^t u_1}{\|u_1\|^2} u_1 = \begin{pmatrix} -51 \\ 167 \\ 24 \end{pmatrix} - \frac{(-51 \ 167 \ 24)\begin{pmatrix} 12 \\ 6 \\ -4 \end{pmatrix}}{12^2 + 6^2 + (-4)^2} \begin{pmatrix} 12 \\ 6 \\ -4 \end{pmatrix}$$

$$= \begin{pmatrix} -69 \\ 158 \\ 30 \end{pmatrix},$$

$$u_3 = a_3 - \text{Proj}_{u_1}(a_3) - \text{Proj}_{u_2}(a_3) = a_3 - \frac{a_3^t u_1}{\|u_1\|^2} u_1 - \frac{a_3^t u_2}{\|u_2\|^2} u_2$$

$$= \begin{pmatrix} 4 \\ -68 \\ -41 \end{pmatrix} - \frac{(4 \ -68 \ -41)\begin{pmatrix} 12 \\ 6 \\ -4 \end{pmatrix}}{196} \begin{pmatrix} 12 \\ 6 \\ -4 \end{pmatrix} - \frac{(4 \ -68 \ -41)\begin{pmatrix} -69 \\ 158 \\ 30 \end{pmatrix}}{30625} \begin{pmatrix} -69 \\ 158 \\ 30 \end{pmatrix}$$

$$= \begin{pmatrix} -58/5 \\ 6/5 \\ -33 \end{pmatrix}.$$

We now find orthonormal vectors:

$$v_1 = \frac{u_1}{\|u_1\|} = \frac{1}{14} \begin{pmatrix} 12 \\ 6 \\ -4 \end{pmatrix},$$

$$v_2 = \frac{u_2}{\|u_2\|} = \frac{1}{175} \begin{pmatrix} -69 \\ 158 \\ 30 \end{pmatrix},$$

$$v_3 = \frac{u_3}{\|u_3\|} = \frac{1}{35} \begin{pmatrix} -58/5 \\ 6/5 \\ -33 \end{pmatrix}.$$

Since $Q = \begin{pmatrix} v_1 & v_2 & v_3 \end{pmatrix}$ then:

$$Q = \begin{pmatrix} 6/7 & -69/175 & -58/175 \\ 3/7 & 158/175 & 6/175 \\ -2/7 & 30/175 & -33/35 \end{pmatrix}.$$

It remains to compute R:

$$R = Q^t A = \begin{pmatrix} 6/7 & 3/7 & -2/7 \\ -69/175 & 158/175 & 30/175 \\ -58/175 & 6/175 & -33/35 \end{pmatrix} \begin{pmatrix} 12 & -51 & 4 \\ 6 & 167 & -68 \\ -4 & 24 & -41 \end{pmatrix}$$

$$= \begin{pmatrix} 14 & 21 & -14 \\ 0 & 175 & -70 \\ 0 & 0 & 35 \end{pmatrix}.$$

13.6 Chapter 6

Exercise 6.1

A set C is convex if for any pair of points $x, y \in C$ and for every $\lambda \in [0, 1]$ it holds that $\lambda x + (1 - \lambda)y \in C$. Now, let X, Y be two left-inverse matrices of A. Show that any convex combination of these matrices yields a left-inverse of A. Conclude that if more than one left-inverse of A exists, then there is an infinite number of left-inverses of A.

Solution. The matrices X, Y are left-inverses of A. We will show that $\lambda X + (1 - \lambda)Y$ is also a left-inverse of A, for every $\lambda \in [0, 1]$. Indeed,

$$(\lambda X + (1 - \lambda)Y)A = \lambda X A + (1 - \lambda)Y A = \lambda I + (1 - \lambda)I = I.$$

That is, if there are two left-inverses of A, then every convex combination of them is also a left-inverse of A, and therefore there is an infinite number of left-inverses.

Exercise 6.2

Let:

$$A = \begin{pmatrix} -3 & 4 & 1 \\ -4 & 6 & 1 \end{pmatrix}, \quad B = \frac{1}{9}\begin{pmatrix} -11 & 7 \\ -10 & 8 \\ 16 & -11 \end{pmatrix}, \quad C = \frac{1}{2}\begin{pmatrix} 0 & 0 \\ -1 & 1 \\ 6 & -4 \end{pmatrix},$$

$$b = \begin{pmatrix} 1 \\ -2 \end{pmatrix}.$$

(a) Show that the matrices B and C are right-inverses of A. Express by B and C solutions to the set of linear equations $Ax = b$.

(b) Show that the difference between the above solutions yields a solution to the homogeneous set of equations $Ax = 0$.

Solution. It is easy to check numerically that $AC = AB = I$. Now, choose

$$x_1 = Cb, \quad x_2 = Bb,$$

and therefore

$$Ax_1 = ACb = Ib = b,$$
$$Ax_2 = ABb = Ib = b.$$

Thus, x_1 and x_2 are the solutions to the above equation. Computing them, we get

$$x_1 = Cb = \frac{1}{2}\begin{pmatrix} 0 & 0 \\ -1 & 1 \\ 6 & -4 \end{pmatrix}\begin{pmatrix} 1 \\ -2 \end{pmatrix} = \begin{pmatrix} 0 \\ -1.5 \\ 7 \end{pmatrix}.$$

Checking this out:

$$Ax_1 = \begin{pmatrix} -3 & 4 & 1 \\ -4 & 6 & 1 \end{pmatrix}\begin{pmatrix} 0 \\ -1.5 \\ 7 \end{pmatrix} = \begin{pmatrix} 1 \\ -2 \end{pmatrix}.$$

Likewise

$$x_2 = Bb = \frac{1}{9}\begin{pmatrix} -11 & 7 \\ -10 & 8 \\ 16 & -11 \end{pmatrix}\begin{pmatrix} 1 \\ -2 \end{pmatrix} = \begin{pmatrix} -25/9 \\ -26/9 \\ 38/9 \end{pmatrix}.$$

Checkout

$$Ax_2 = \begin{pmatrix} -3 & 4 & 1 \\ -4 & 6 & 1 \end{pmatrix}\begin{pmatrix} -25/9 \\ -26/9 \\ 38/9 \end{pmatrix} = \begin{pmatrix} 1 \\ -2 \end{pmatrix}.$$

(b) It can be easily seen that

$$A(x_1 - x_2) = Ax_1 - Ax_2 = b - b = 0.$$

Exercise 6.3

Prove theorem 7.3.

Solution. Assume that $A \in R^{n \times n}$. Let us suppose first that the system of equations $Ax = b$ has a solution for any b. Since $b \in R^n$, and Ax can be regarded as a linear combination of the columns of A, then we can say that the n columns of A span R^n, which implies that the columns of A are linearly independent (since the number of columns coincides with the dimension of R^n). Thus, by (2.1) (item 3) the system of equations $Ax = b$ has a unique solution for any b. For the second part, suppose for some right-hand side b, say b' there exists a unique solution such that $Ax = b'$, say $Ax' = b'$. Consider next the system of linear equation $Ax = \underline{0}$. Had it have a non-trivial solution $Ay = \underline{0}$, this would lead to $A(x' + y) = b'$, violating the uniqueness assumption of x'. Thus, $Ax = \underline{0}$ comes only with the trivial solution. This implies that the columns of A are linearly independent and since there are n of them, they form a basis. Thus, any vector $b \in R^n$ can be expressed a unique linear combination of these columns. In particular, $Ax = b$ has a unique solution for any $b \in R^n$, as required.

Exercise 6.4

Consider the following matrix:

$$A = \begin{pmatrix} 12 & -51 & 4 \\ 6 & 167 & -68 \\ -4 & 24 & -41 \end{pmatrix}.$$

Given its QR factorization as computed in Exercise 5.9, compute its inverse: A^{-1}.

Solution. Using the results of Exercise 5.9, the QR factorization of A is given as:

$$Q = \begin{pmatrix} 6/7 & -69/175 & -58/175 \\ 3/7 & 158/175 & 6/175 \\ -2/7 & 30/175 & -33/35 \end{pmatrix}, \quad R = \begin{pmatrix} 14 & 21 & -14 \\ 0 & 175 & -70 \\ 0 & 0 & 35 \end{pmatrix}.$$

Since $A^{-1} = R^{-1}Q^t$, we need to find first the inverse of R. We do that using backward substitution. Computing the first row of R^{-1}:

$$\begin{pmatrix} 14 & 21 & -14 \\ 0 & 175 & -70 \\ 0 & 0 & 35 \end{pmatrix} \begin{pmatrix} x_1 \\ x_2 \\ x_3 \end{pmatrix} = \begin{pmatrix} 1 \\ 0 \\ 0 \end{pmatrix} \implies \begin{cases} x_3 = x_2 = 0, \\ 14x_1 = 1 \implies x_1 = 1/14. \end{cases}$$

Second row of R^{-1}:

$$\begin{pmatrix} 14 & 21 & -14 \\ 0 & 175 & -70 \\ 0 & 0 & 35 \end{pmatrix} \begin{pmatrix} x_1 \\ x_2 \\ x_3 \end{pmatrix} = \begin{pmatrix} 0 \\ 1 \\ 0 \end{pmatrix} \implies \begin{cases} x_3 = 0 \\ 175x_2 = 1 \implies x_2 = 1/175 \\ 14x_1 + 21 \cdot \frac{1}{175} = 0 \implies x_1 = -3/350. \end{cases}$$

Third row of R^{-1}:

$$\begin{pmatrix} 14 & 21 & -14 \\ 0 & 175 & -70 \\ 0 & 0 & 35 \end{pmatrix} \begin{pmatrix} x_1 \\ x_2 \\ x_3 \end{pmatrix} = \begin{pmatrix} 0 \\ 0 \\ 1 \end{pmatrix} \implies \begin{cases} x_3 = 1/35 \\ 175x_2 - 70\frac{1}{35} = 0 \implies x_2 = 2/175 \\ 14x_1 + 21\frac{2}{175} - 14\frac{1}{35} = 0 \implies x_1 = 2/175. \end{cases}$$

Altogether, we get:

$$R^{-1} = \begin{pmatrix} 1/14 & -3/350 & 2/175 \\ 0 & 1/175 & 2/175 \\ 0 & 0 & 1/35 \end{pmatrix},$$

and therefore:

$$A^{-1} = R^{-1}Q^t = \begin{pmatrix} 1/14 & -3/350 & 2/175 \\ 0 & 1/175 & 2/175 \\ 0 & 0 & 1/35 \end{pmatrix} \begin{pmatrix} 6/7 & 3/7 & -2/7 \\ -69/175 & 158/175 & 30/175 \\ -58/175 & 6/175 & -33/35 \end{pmatrix}$$

$$= \begin{pmatrix} 149/2450 & 57/2450 & -8/245 \\ -37/6125 & 34/6125 & -12/1225 \\ -58/6125 & 6/6125 & -33/1225 \end{pmatrix}.$$

Exercise 6.5

Let

$$A = \begin{pmatrix} 0 & 1 & 2 \\ 1 & 0 & 3 \\ 4 & -3 & 8 \end{pmatrix}, \quad b = \begin{pmatrix} 1 \\ 1 \\ 1 \end{pmatrix}.$$

(a) Solve the system of equations $Ax = b$ by transforming to the system $Lx = b'$, where L is an upper triangular matrix, using row operations on A.

(b) Solve the same system of linear equations ($Ax = b$) by transforming to $Ix = b''$, using row operations on L.

(c) Find A^{-1} using row operations on A and the same operation to I.

Solution. (a)

$$\begin{pmatrix} 0 & 1 & 2 & | & 1 \\ 1 & 0 & 3 & | & 1 \\ 4 & -3 & 8 & | & 1 \end{pmatrix} \xrightarrow{R_1 \leftrightarrow R_2} \begin{pmatrix} 1 & 0 & 3 & | & 1 \\ 0 & 1 & 2 & | & 1 \\ 4 & -3 & 8 & | & 1 \end{pmatrix} \xrightarrow{R_3 \leftarrow -4R_1 + R_3} \begin{pmatrix} 1 & 0 & 3 & | & 1 \\ 0 & 1 & 2 & | & 1 \\ 0 & -3 & -4 & | & -3 \end{pmatrix}$$

$$\xrightarrow{R_3 \leftarrow 3R_2 + R_3} \begin{pmatrix} 1 & 0 & 3 & | & 1 \\ 0 & 1 & 2 & | & 1 \\ 0 & 0 & 2 & | & 0 \end{pmatrix},$$

and we obtained $Lx = b'$, i.e $\begin{pmatrix} 1 & 0 & 3 \\ 0 & 1 & 2 \\ 0 & 0 & 2 \end{pmatrix} \begin{pmatrix} x_1 \\ x_2 \\ x_3 \end{pmatrix} = \begin{pmatrix} 1 \\ 0 \\ 0 \end{pmatrix}$, whose solution is obtained through:

$$2x_3 = 0 \implies x_3 = 0,$$
$$x_2 + 2x_3 = 1 \implies x_2 = 1,$$
$$x_1 + 3x_3 = 1 \implies x_1 = 1.$$

(b)

$$\begin{pmatrix} 1 & 0 & 3 & | & 1 \\ 0 & 1 & 2 & | & 1 \\ 0 & 0 & 2 & | & 0 \end{pmatrix} \xrightarrow[R_3 \leftarrow \frac{1}{2} R_3]{R_2 \leftarrow R_2 - R_3} \begin{pmatrix} 1 & 0 & 3 & | & 1 \\ 0 & 1 & 0 & | & 1 \\ 0 & 0 & 1 & | & 0 \end{pmatrix} \xrightarrow{R_1 \leftarrow R_1 - 3R_3} \begin{pmatrix} 1 & 0 & 0 & | & 1 \\ 0 & 1 & 0 & | & 1 \\ 0 & 0 & 1 & | & 0 \end{pmatrix}$$

and we got: $x_1 = x_2 = 1$, $x_3 = 0$.

(c) We perform the same transformations as previously, now on the following system:

$$\begin{pmatrix} 0 & 1 & 2 & | & 1 & 0 & 0 \\ 1 & 0 & 3 & | & 0 & 1 & 0 \\ 4 & -3 & 8 & | & 0 & 0 & 1 \end{pmatrix}$$

$$\xrightarrow{R_1 \leftrightarrow R_2} \begin{pmatrix} 1 & 0 & 3 & | & 0 & 1 & 0 \\ 0 & 1 & 2 & | & 1 & 0 & 0 \\ 4 & -3 & 8 & | & 0 & 0 & 1 \end{pmatrix} \xrightarrow{R_3 \leftarrow -4R_1 + R_3} \begin{pmatrix} 1 & 0 & 3 & | & 0 & 1 & 0 \\ 0 & 1 & 2 & | & 1 & 0 & 0 \\ 0 & -3 & -4 & | & 0 & -4 & 1 \end{pmatrix}$$

$$\xrightarrow{R_3 \leftarrow 3R_2 + R_3} \begin{pmatrix} 1 & 0 & 3 & | & 0 & 1 & 0 \\ 0 & 1 & 2 & | & 1 & 0 & 0 \\ 0 & 0 & 2 & | & 3 & -4 & 1 \end{pmatrix} \xrightarrow[R_3 \leftarrow \frac{1}{2} R_3]{R_2 \leftarrow R_2 - R_3} \begin{pmatrix} 1 & 0 & 3 & | & 0 & 1 & 0 \\ 0 & 1 & 0 & | & -2 & 4 & -1 \\ 0 & 0 & 1 & | & 3/2 & -2 & 1/2 \end{pmatrix}$$

$$\xrightarrow{R_1 \leftarrow R_1 - 3R_3} \begin{pmatrix} 1 & 0 & 0 & | & -9/2 & 7 & -3/2 \\ 0 & 1 & 0 & | & -2 & 4 & -1 \\ 0 & 0 & 1 & | & 3/2 & -2 & 1/2 \end{pmatrix},$$

and therefore

$$A^{-1} = \begin{pmatrix} -9/2 & 7 & -3/2 \\ -2 & 4 & -1 \\ 3/2 & -2 & 1/2 \end{pmatrix}.$$

Exercise 6.6

Show that if E is a row operation matrix then its inverse exists. Specifically,

1. if E multiplies all entries of row i by a $c \neq 0$, then E^{-1} does the same but with $1/c$,
2. if E swaps two rows, then $E^{-1} = E$, and
3. if E subtracts c times row j from row i, then E^{-1} adds c times row j to row i.

Conclude that the inverse, as expected, cancels the row operation originally made.

Solution. 1. Such E has the following structure:

$$
E = \begin{pmatrix} 1 & & & & \\ & \ddots & & & \\ & & c & & \\ & & & \ddots & \\ & & & & 1 \end{pmatrix},
$$

where $c \neq 0$ is in the ith row. It is easy to see that E^{-1} is the same matrix as E, only the diagonal at the ith row is $1/c$.

2. The corresponding matrix E is obtained by swapping the ith and jth row of the identity matrix. By standard computation, it can be seen that if E is multiplied by the same matrix E then we get the identity matrix again.

3. The corresponding matrix E is the same as the identity matrix, except from the i, jth entry, whose value is $-c$. It can be seen then that the matrix which is identical to E, except the i, j-entry whose value is c, is its inverse; multiplying row i with column j yields $-c + c = 0$, which is the i, jth (non-diagonal) entry. All other non-diagonal entries are zero, and all diagonal entries are 1.

Exercise 6.7

(a) Show that if A, B are similar then also $A - aI$ and $B - aI$ are similar for any $a \in R$.
(b) Let $A, B \in R^{n \times n}$ be similar matrices. Prove that $trace(A) = trace(B)$. Use the fact that $trace(CD) = trace(DC)$ for any square matrices C and D (both of the same dimension).

(c) Prove that the diagonal matrices

$$D_1 = \begin{pmatrix} a & 0 & 0 \\ 0 & b & 0 \\ 0 & 0 & c \end{pmatrix}, \quad D_2 = \begin{pmatrix} c & 0 & 0 \\ 0 & a & 0 \\ 0 & 0 & b \end{pmatrix}$$

are similar. What is S that satisfies $D_2 = SD_1S^{-1}$?

Solution. (a) Since A, B are similar, a square matrix P exists such that $A = P^{-1}BP$. This implies that

$$A - aI = P^{-1}BP - aI = P^{-1}BP - aP^{-1}P$$
$$= P^{-1}BP - aP^{-1}IP = P^{-1}(B - aI)P.$$

(b) There exists an invertible matrix P such that $A = P^{-1}BP$. Thus,

$$trace(A) = trace(P^{-1}BP) = trace(BP^{-1}P) = trace(B),$$

using $trace(CD) = trace(DC)$ in the second equation.

(c) We note that D_2 can be obtained by re-arranging the diagonal entries of D_1, or in other words, performing elementary row and column operations on D_1. We choose:

$$S = \begin{pmatrix} 0 & 0 & 1 \\ 1 & 0 & 0 \\ 0 & 1 & 0 \end{pmatrix} \implies S^{-1} = \begin{pmatrix} 0 & 1 & 0 \\ 0 & 0 & 1 \\ 1 & 0 & 0 \end{pmatrix},$$

therefore,

$$SD_1S^{-1} = \begin{pmatrix} 0 & 0 & 1 \\ 1 & 0 & 0 \\ 0 & 1 & 0 \end{pmatrix} \begin{pmatrix} a & 0 & 0 \\ 0 & b & 0 \\ 0 & 0 & c \end{pmatrix} \begin{pmatrix} 0 & 1 & 0 \\ 0 & 0 & 1 \\ 1 & 0 & 0 \end{pmatrix} = \begin{pmatrix} 0 & 0 & c \\ a & 0 & 0 \\ 0 & b & 0 \end{pmatrix} \begin{pmatrix} 0 & 1 & 0 \\ 0 & 0 & 1 \\ 1 & 0 & 0 \end{pmatrix} = \begin{pmatrix} c & 0 & 0 \\ 0 & a & 0 \\ 0 & 0 & b \end{pmatrix} = D_2.$$

Exercise 6.8

Let $A = \begin{pmatrix} 1 & 2 \\ 3 & 4 \end{pmatrix}$ be the transformation matrix from R^2 to R^2 with respect to the standard basis. Let $\begin{pmatrix} 2 \\ 5 \end{pmatrix}$, $\begin{pmatrix} 1 \\ 3 \end{pmatrix}$ be another basis.

(a) Show that the transformation matrix with respect to that basis is $B = \begin{pmatrix} -5 & -8 \\ 6 & 10 \end{pmatrix}$.

(b) Check your result: let $S = \begin{pmatrix} 1 & 2 \\ 3 & 5 \end{pmatrix}$. Compute S^{-1} and verify that $A = SBS^{-1}$.

Solution. (a) We define $S = \left(\begin{smallmatrix}1 & 2\\ 3 & 5\end{smallmatrix}\right)$. We then can compute $B = S^{-1}AS$.

(b) Since S is a 2×2 matrix, we can compute the inverse as follows:

$$S^{-1} = \frac{1}{5-6}\begin{pmatrix} 5 & -2 \\ -3 & 1 \end{pmatrix} = \begin{pmatrix} -5 & 2 \\ 3 & -1 \end{pmatrix},$$

and therefore,

$$A = SBS^{-1} = \begin{pmatrix} 1 & 2 \\ 3 & 5 \end{pmatrix}\begin{pmatrix} -5 & -8 \\ 6 & 10 \end{pmatrix}\begin{pmatrix} -5 & 2 \\ 3 & -1 \end{pmatrix}$$

$$= \begin{pmatrix} 7 & 12 \\ 15 & 26 \end{pmatrix}\begin{pmatrix} -5 & 2 \\ 3 & -1 \end{pmatrix} = \begin{pmatrix} 1 & 2 \\ 3 & 4 \end{pmatrix}.$$

Exercise 6.9

Let $A \in R^{n \times n}$ be a square matrix. Let $E_1, \ldots, E_m$ and $F_1, \ldots, F_k$ two sequences of row operation matrices. Suppose that

$$E_m \cdots E_1 A F_1 \cdots F_k = I$$

(a) Explain why A is invertible.
(b) Express A^{-1} in terms of the above row operation matrices.

Solution. (a) Note that the matrices at both sides of the equations are fully ranked, since the one at the right-hand side is the identity matrix. Since row and column operations preserve the rank, then A must be fully ranked, and therefore invertible.

(b) We can multiply both sides of the equation by $(AF_1 \cdots F_k)^{-1} = F_k^{-1} \cdots F_1^{-1}A^{-1}$ from the right to obtain:

$$E_m \cdots E_1 = F_k^{-1} \cdots F_1^{-1}A^{-1}.$$

To isolate A^{-1}, we multiply both sides by $F_1 \cdots F_k$ to obtain:

$$F_1 \cdots F_k E_m \cdots E_1 = A^{-1}.$$

Exercise 6.10

Let $T : R^2 \to R^2$ be a linear transformation defined as

$$T(x, y) = \begin{pmatrix} 3x + 4y \\ 2x - 5y \end{pmatrix}.$$

Given the following bases:

$$E = \{e_1, e_2\}; \quad S = \left\{ \begin{pmatrix} 1 \\ 2 \end{pmatrix}, \begin{pmatrix} 2 \\ 3 \end{pmatrix} \right\}.$$

(a) Find the transformation matrix A which represents T with respect to E, in both domain and range.

(b) Find the transformation matrix A which represents T with respect to S, in both domain and range.

(c) What is the transformation matrix if the domain subspace is expressed in terms of S and the range subspace is expressed in terms of $S = \left\{ \begin{pmatrix} -1 \\ 1 \end{pmatrix}, \begin{pmatrix} 2 \\ -1 \end{pmatrix} \right\}$?

Solution. (a) The transformation matrix is given as

$$A = (T(1,0), T(0,1)) = \begin{pmatrix} 3 & 4 \\ 2 & -5 \end{pmatrix}.$$

(b) Denote $C = \begin{pmatrix} 1 & 2 \\ 2 & 3 \end{pmatrix}$. The transformation matrix then becomes

$$C^{-1}AC = \begin{pmatrix} -49 & -76 \\ 30 & 47 \end{pmatrix}.$$

(c) Denote

$$B = \begin{pmatrix} -1 & 2 \\ 1 & -1 \end{pmatrix}.$$

We showed in this case that

$$B^{-1}AC = \begin{pmatrix} -5 & -4 \\ 3 & 7 \end{pmatrix}.$$

13.7 Chapter 7

Exercise 7.1

Denote $AA^\dagger = E$.

(a) Show that for every $k \in N$ it holds that $E^k = E$.

(b) Show that $(I - E)E = E(I - E) = 0$.

(c) Show that E is symmetric.

Solution. (a) We prove that by induction. For $k = 2$ we have:

$$E^2 = (AA^\dagger)^2 = A(A^t A)^{-1} A^t A(A^t A)^{-1} A^t = A(A^t A)^{-1} A^t.$$

Suppose now that $E^m = E$ holds for $m = k - 1$. We prove now that it holds for $m = k$ as well:

$$E^k = (AA^\dagger)^k = (AA^\dagger)^{k-1} AA^\dagger = (A(A^t A)^{-1} A^t)^{k-1} A(A^t A)^{-1} A^t$$
$$= A(A^t A)^{-1} A^t A(A^t A)^{-1} A^t = A(A^t A)^{-1} A^t = E.$$

(b)

$$E(I - E) = E - E^2 = E - E = 0,$$
$$(I - E)E = E - E^2 = E - E = 0.$$

(c)

$$E^t = (AA^\dagger)^t = (A(A^t A)^{-1} A^t)^t = A[(A^t A)^{-1}]^t A^t$$
$$= (A(A^t A)^{-1} A^t)^t = A(A^t A)^{-1} A^t = E.$$

Exercise 7.2

Assume $A \in R^{m \times n}$ with $m \geq n$ is a full-rank matrix. Show that $A^\dagger \in R^{n \times m}$ uniquely obeys for $X \in R^{n \times m}$ the following four conditions: $AXA = A$, $XAX = X$, $(AX)^t = AX$ and $(XA)^t = XA$.

Solution. First, we observe that if both X and Y obey these conditions then $AY = AX$. Indeed,

$$AY = (AY)^t = (AXAY)^t = (AY)^t(AX)^t = AYAX = AX.$$

In the same fashion it is possible to argue that $YA = XA$. Then,

$$Y = YAY = XAY = XAX = X,$$

completing the uniqueness part of the proof. Second,

$$AA^\dagger A = A(A^t A)^{-1} A^t A = AI = A,$$
$$A^\dagger AA^\dagger = (A^t A)^{-1} A^t AA^\dagger = IA^\dagger = A^\dagger.$$

In Exercise 7.1, we showed that $(AA^\dagger)^t = AA^\dagger$.

Finally, since $A^\dagger$ is a left-inverse of A, $A^\dagger A = I$, which of course is a symmetric matrix.

Exercise 7.3

Let $Q^{m \times n}$ be an orthonormal matrix with $n \le m$.

(a) Show that $\sum_{i=1}^{n} q_i q_i^t = QQ^t$, where q_i is the ith column of Q.

(b) Assume that $A \in R^{m \times n}$, $m \ge n$ and that the columns of A are linearly independent. Show that $AA^\dagger = QQ^t$, where Q is the first term in the QR factorization of A.

(c) Conclude that $\|QQ^t b - b\| = \min_x \|Ax - b\|$.

(d) Show that QQ^t is the projection matrix of the subspace spanned by A's columns, and that $I - QQ^t$ is the residual matrix.

Solution. (a) It holds that

$$\left[\sum_{i=1}^{n} q_i q_i^t \right]_{jk} = \sum_{i=1}^{n} [q_i q_i^t]_{jk} = \sum_{i=1}^{n} [q_i]_j [q_i^t]_k.$$

On the other hand,

$$[QQ^t]_{jk} = \sum_{i=1}^{n} Q_{ji} Q_{ik}^t = \sum_{i=1}^{n} Q_{ji} Q_{ki} = \sum_{i=1}^{n} [q_i]_j [q_i]_k,$$

and we got that for every j, k

$$\left[\sum_{i=1}^{n} q_i q_i^t \right]_{jk} = \left[QQ^t \right]_{jk}.$$

(b)

$$AA^\dagger = A(A^t A)^{-1} A^t = QR((QR)^t QR)^{-1}(QR)^t = QR(R^t Q^t QR)^{-1} R^t Q^t$$
$$= QR(R^t R)^{-1} R^t Q^t \stackrel{*}{=} QRR^{-1}(R^t)^{-1} R^t Q^t = QQ^t$$

* for invertible A, B it holds that $(AB)^{-1} = B^{-1} A^{-1}$.

(c) We know that the solution for $\min_x \|Ax - b\|$ is $x^* = A^\dagger b = (A^t A)^{-1} A^t b$, and therefore $Ax^* = A(A^t A)^{-1} A^t b$. Moreover, as shown previously, $AA^\dagger b = QQ^t b$, and therefore $\min_x \|Ax - b\| = \|QQ^t b - b\|$.

(d) From previous results we know that $AA^\dagger = QQ^t$ is a projection matrix. The residual matrix is therefore $I - QQ^t$.

Exercise 7.4

Let

$$A = \begin{pmatrix} -3 & -4 \\ 4 & 6 \\ 1 & 1 \end{pmatrix}, \quad b = \begin{pmatrix} 1 \\ -2 \\ 0 \end{pmatrix}.$$

(a) Compute the QR factorization of A.
(b) Compute $A^\dagger$.
(c) Check if there exists a solution to $Ax = b$, and if so, then $x = A^\dagger b$ is the unique solution.

Solution. (a) We obtain the matrix Q by applying the Gram–Schmidt process on the columns of A

$$Q = \begin{pmatrix} -0.5883 & -0.4576 \\ 0.7845 & -0.5230 \\ 0.1961 & 0.7191 \end{pmatrix}.$$

We then obtain the matrix R through $R = Q^t A$

$$R = \begin{pmatrix} 5.099 & 7.2563 \\ 0 & -0.5883 \end{pmatrix}.$$

(b) We note that the matrix $A^t A$ is a 2×2 matrix, and so we can use the closed form solution to matrix inversion. Altogether

$$A^\dagger = \begin{pmatrix} -1.222 & -1.11 & 1.778 \\ 0.778 & 0.889 & -1.22 \end{pmatrix}.$$

(c) We compute

$$x = A^\dagger b = \begin{pmatrix} -1.222 & -1.11 & 1.778 \\ 0.778 & 0.889 & -1.22 \end{pmatrix} \begin{pmatrix} 1 \\ -2 \\ 0 \end{pmatrix} = \begin{pmatrix} 1 \\ -1 \end{pmatrix},$$

and check

$$Ax = \begin{pmatrix} -3 & -4 \\ 4 & 6 \\ 1 & 1 \end{pmatrix} \begin{pmatrix} 1 \\ -1 \end{pmatrix} = \begin{pmatrix} 1 \\ -2 \\ 0 \end{pmatrix} = b.$$

The equality holds and thus we found a unique solution to $Ax = b$.

13.8   Chapter 8

Exercise 8.1

For the following permutations:

$$\sigma_1 = \begin{pmatrix} 1 & 2 & 3 & 4 & 5 \\ 2 & 4 & 5 & 1 & 3 \end{pmatrix}, \quad \sigma_2 = \begin{pmatrix} 1 & 2 & 3 & 4 & 5 \\ 4 & 1 & 3 & 5 & 2 \end{pmatrix}.$$

(a) Compute σ^{-1} and $\sigma_1 \circ \sigma_2$.
(b) For every permutation, determine its sign (including permutations computed in (a)).
(c) Show that in general $sgn(\sigma) = sgn(\sigma^{-1})$.

Solution. (a)

$$\sigma_1^{-1} = \begin{pmatrix} 1 & 2 & 3 & 4 & 5 \\ 4 & 1 & 5 & 2 & 3 \end{pmatrix},$$

$$\sigma_1 \circ \sigma_2 = \begin{pmatrix} 1 & 2 & 3 & 4 & 5 \\ 1 & 2 & 5 & 3 & 4 \end{pmatrix}.$$

(b)

$$sgn(\sigma_1) = -1 \implies sgn(\sigma_1^{-1}) = -1,$$
$$sgn(\sigma_2) = -1,$$
$$sgn(\sigma_1 \circ \sigma_2) = sgn(\sigma_1)sgn(\sigma_2) = 1.$$

(c) We know that $sgn(\sigma \circ \sigma^{-1}) = sgn(Id) = 1$, and therefore $sgn(\sigma \circ \sigma^{-1}) = sgn(\sigma)sgn(\sigma^{-1}) = 1$. Thus, either both terms in the product equal to 1 or -1. In either case, we get $sgn(\sigma) = sgn(\sigma^{-1})$.

Exercise 8.2

Compute the determinants of the following matrices:

$$A = \begin{pmatrix} 2 & -3 \\ 4 & 7 \end{pmatrix}, \quad B = \begin{pmatrix} 1 & -2 & 3 \\ 2 & 4 & -1 \\ 1 & 5 & -2 \end{pmatrix}, \quad C = \begin{pmatrix} 1/2 & -1 & -1/3 \\ 3/4 & 1/2 & -1 \\ 1 & -4 & 1 \end{pmatrix}.$$

Solution. Using the formula for a 2×2 determinant, we get

$$det(A) = 2 \cdot 7 - (-3) \cdot 4 = 26,$$

$$det(B) = 1 \cdot det \left(\begin{smallmatrix} 4 & -1 \\ 5 & -2 \end{smallmatrix} \right) + 2 \cdot det \left(\begin{smallmatrix} 2 & -1 \\ 1 & -2 \end{smallmatrix} \right) + 3 \cdot det \left(\begin{smallmatrix} 2 & 4 \\ 1 & 5 \end{smallmatrix} \right) = 9.$$

Regarding matrix C, we multiply its first row by 6 and its second row by 4, to get rid of the fractions. Denote the new matrix by C', we use the property proved in exercise ??:

$$det(C') = 24 det(C).$$

Compute then

$$det(C') = det \left(\begin{smallmatrix} 3 & -6 & -2 \\ 3 & 2 & -4 \\ 1 & -4 & 1 \end{smallmatrix} \right) = 28,$$

and therefore

$$det(C) = \frac{1}{24} det(C') = \frac{7}{6}.$$

Exercise 8.3

Claim: for $A \in R^{n \times n}$, we have $det(A) = det(A^t)$. Prove that claim for matrices of dimension 2×2.

Solution.

$$A = \begin{pmatrix} a_{11} & a_{12} \\ a_{21} & a_{22} \end{pmatrix}, \quad A^t = \begin{pmatrix} a_{11} & a_{21} \\ a_{12} & a_{22} \end{pmatrix}$$

then

$$det(A) = a_{11}a_{22} - a_{12}a_{21} = a_{11}a_{22} - a_{21}a_{12} = det(A^t).$$

Exercise 8.4

Prove theorem equal rows using the three row/column operations and their effect on the determinant.

Solution. We know that if we swap two rows of a matrix A, then the resulting matrix A' satisfies $det(A') = -det(A)$. On the other hand, if we swap two identical rows then the matrix A remains unchanged, and therefore $det(A') = det(A)$. Altogether we get $det(A) = 0$. The same holds for column operations.

Exercise 8.5

(a) Let $A \in R^{n \times n}$ and let A' be the matrix obtained by multiplying the ith row of matrix A by a scalar λ. Prove that $det(A') = \lambda det(A)$.

(b) Let $A \in R^{n \times n}$ and let A' be the matrix obtained by swapping two rows of matrix A. Prove that $det(A') = -det(A)$.

Solution. (a) Note that for $i \neq j$ it holds that $a'_{j\sigma(j)} = a_{j\sigma(j)}$, and for $j = i$ it holds that $a'_{j\sigma(j)} = \lambda a_{j\sigma(j)}$. Therefore,

$$det(A') = \sum_{\sigma \in S_n} sgn(\sigma) \prod_{j=1}^{n} a'_{j\sigma(j)} = \sum_{\sigma \in S_n} sgn(\sigma) a'_{i\sigma(i)} \prod_{j=1, j\neq i}^{n} a_{j\sigma(j)}$$

$$= \sum_{\sigma \in S_n} sgn(\sigma) \lambda a_{i\sigma(i)} \prod_{j=1, j\neq i}^{n} a_{j\sigma(j)} = \lambda \sum_{\sigma \in S_n} sgn(\sigma) \prod_{j=1}^{n} a_{j\sigma(j)}$$

$$= \lambda det(A).$$

(b) Let us compute the determinants $det(A)$ and $det(A')$ according to the definition

$$det(A) = \sum_{\sigma \in S_n} sgn(\sigma) \prod_{k=1}^{n} a_{k\sigma(k)},$$

$$det(A') = \sum_{\hat{\sigma} \in S_n} sgn(\hat{\sigma}) \prod_{k=1}^{n} a_{k\hat{\sigma}(k)}.$$

where for every $\sigma \in S_n$, the permutation $\hat{\sigma}$ equals to

$$\hat{\sigma}(k) = \begin{cases} \sigma(k) & k \neq i, j, \\ \sigma(i) & k = j, \\ \sigma(j) & k = i. \end{cases}$$

That is, the permutation $\hat{\sigma}$ is exactly like σ, except the swap between i and j. It is easy to see that $\hat{\sigma}$ changes signs; from an even number of pairs k_1, k_2 such that $\sigma(k_1) < \sigma(k_2)$ to an odd number, or vice versa. Thus, the sign of the whole determinant changes:

$$det(A') = \sum_{\hat{\sigma} \in S_n} sgn(\hat{\sigma}) \prod_{k=1}^{n} a_{k\hat{\sigma}(k)}$$

$$= \sum_{\sigma \in S_n} -sgn(\sigma) a_{i\sigma(j)} a_{j\sigma(i)} \prod_{k \neq i,j}^{n} a_{k\sigma(k)} = - \sum_{\sigma \in S_n} sgn(\sigma) \prod_{k=1}^{n} a_{k\sigma(k)}$$

$$= -det(A).$$

Exercise 8.6

Let $A, B, C \in R^{n \times n}$. Suppose that A is identical to B, except the i^*th column, and suppose that C is identical to A, but its i^*th column is the sum $A_{.i^*} + B_{.i^*}$. Find an expression for $det(C)$ in terms of $det(A)$ and $det(B)$.

Solution.

$$det(C) = det(C^t) = \sum_{\sigma \in S_n} sgn(\sigma) \prod_{i=1}^{n} c_{i\sigma(i)} = \sum_{\sigma \in S_n} sgn(\sigma) c_{i^*\sigma(i^*)} \prod_{i \neq i^*}^{n} c_{i\sigma(i)}$$

$$= \sum_{\sigma \in S_n} sgn(\sigma)(a_{i^*\sigma(i^*)} + b_{i^*\sigma(i^*)}) \prod_{i \neq i^*}^{n} c_{i\sigma(i)}$$

$$= \sum_{\sigma \in S_n} sgn(\sigma) a_{i^*\sigma(i^*)} \prod_{i \neq i^*}^{n} c_{i\sigma(i)} + \sum_{\sigma \in S_n} sgn(\sigma) b_{i^*\sigma(i^*)} \prod_{i \neq i^*}^{n} c_{i\sigma(i)}$$

$$= \sum_{\sigma \in S_n} sgn(\sigma) a_{i^*\sigma(i^*)} \prod_{i \neq i^*}^{n} a_{i\sigma(i)} + \sum_{\sigma \in S_n} sgn(\sigma) b_{i^*\sigma(i^*)} \prod_{i \neq i^*}^{n} b_{i\sigma(i)}$$

$$= \sum_{\sigma \in S_n} sgn(\sigma) \prod_{i=1}^{n} a_{i\sigma(i)} + \sum_{\sigma \in S_n} sgn(\sigma) \prod_{i=1}^{n} b_{i\sigma(i)} = det(A) + det(B).$$

Exercise 8.7

Compute the determinant of the following matrices:

$$A = \begin{pmatrix} 2 & 0 & 1 \\ 2 & 0 & 5 \\ 3 & 7 & 2 \end{pmatrix}, \quad B = \begin{pmatrix} 5 & 7 & 0 & 1 \\ 3 & 1 & 0 & 3 \\ 1 & 1 & 1 & 2 \\ 3 & 0 & 0 & 4 \end{pmatrix}, \quad C = \begin{pmatrix} 0 & 1 & 0 & 0 \\ 1 & 0 & 1 & 1 \\ 1 & 1 & -1 & 0 \\ 1 & 1 & 0 & -1 \end{pmatrix}.$$

Solution. We develop the determinant of A through its second column

$$det(A) = \sum_{i=1}^{3}(-1)^{i+2}a_{i2}M_{i2} = -1 \cdot 7 \cdot det\left(\begin{smallmatrix} 2 & 1 \\ 2 & 5 \end{smallmatrix}\right) = -56.$$

Developing $det(B)$ through its fourth row, we get

$$det(B) = \sum_{j=1}^{4}(-1)^{4+j}b_{4j}M_{4j}$$

$$= -1 \cdot 3 \cdot det\left(\begin{smallmatrix} 7 & 0 & 1 \\ 1 & 0 & 3 \\ 1 & 1 & 2 \end{smallmatrix}\right) + 4 \cdot det\left(\begin{smallmatrix} 5 & 7 & 0 \\ 3 & 1 & 0 \\ 1 & 1 & 1 \end{smallmatrix}\right) = 60 - 64 = -4.$$

We computed the minors using the columns who are the most sparse. Finally, we compute $det(C)$ using the first row:

$$det(C) = (-1)^{1+2}M_{12} = -det\left(\begin{smallmatrix} 1 & 1 & 1 \\ 1 & -1 & 0 \\ 1 & 0 & -1 \end{smallmatrix}\right) = -3.$$

Exercise 8.8

Show that

$$det \begin{pmatrix} 1 & 1 & 1 & \cdots & 1 \\ 1 & 2-x & 1 & \cdots & 1 \\ 1 & 1 & 3-x & \cdots & 1 \\ \vdots & \vdots & \vdots & \ddots & \vdots \\ 1 & 1 & 1 & \cdots & n+1-x \end{pmatrix} = \prod_{i=1}^{n}(i-x).$$

Hint: Use row/column operations to obtain a triangular matrix.

Solution. For every $2 \leq i \leq n$ we do $R_i \leftarrow R_i - R_1$. The determinant is invariant to these row operations. Thus,

$$
det \begin{pmatrix} 1 & 1 & 1 & \cdots & 1 \\ 1 & 2-x & 1 & \cdots & 1 \\ 1 & 1 & 3-x & \cdots & 1 \\ \vdots & \vdots & \vdots & \ddots & \vdots \\ 1 & 1 & 1 & \cdots & n+1-x \end{pmatrix}
$$

$$
= det \begin{pmatrix} 1 & 1 & 1 & \cdots & 1 \\ 0 & 1-x & 0 & \cdots & 0 \\ 0 & 0 & 2-x & \cdots & 0 \\ \vdots & \vdots & \vdots & \ddots & \vdots \\ 0 & 0 & 0 & \cdots & n-x \end{pmatrix} = \prod_{i=1}^{n}(i-x).
$$

Exercise 8.9

Show that

$$
det \begin{pmatrix} sin^2\alpha & 1 & cos^2\alpha \\ sin^2\beta & 1 & cos^2\beta \\ sin^2\gamma & 1 & cos^2\gamma \end{pmatrix} = 0
$$

using the fact that $sin^2\alpha + cos^2\alpha = 1$.

Solution. Clearly, the columns are linearly dependent, since the middle column can be written as the sum of the first and third column. The determinant therefore equals to zero.

Exercise 8.10

Let $A \in R^{n \times n}$.

(a) Prove that for any scalar c, $det(cA) = c^n det(A)$.
(b) Prove that $det(adj(A)) = det(A)^{n-1}$

Solution. (a) We know that multiplying a row of A by a scalar c, the determinant of the resulting matrix A' will satisfy $det(A') = c \cdot det(A)$. Since cA is equivalent to multiplying each row of A by c, with the total of n rows, we get that $det(cA) = c^n det(A)$.

(b) It holds that $A adj(A) = det(A)I$. Therefore,

$$det(A adj(A)) = det(det(A)I) \iff det(A) \cdot det(adj(A))$$
$$= det(A)^n det(I) \iff det(A) \cdot det(adj(A)) = det(A)^n.$$

Thus, from the right-hand side equation we get that if $det(A) \neq 0$ then $det(adj(A)) = det(A)^{n-1}$, and if $det(A) = 0$ then $det(adj(A)) = 0$. In any case, we proved that $det(adj(A)) = det(A)^{n-1}$.

Exercise 8.11

Let

$$A = \begin{pmatrix} 2 & 3 & 4 \\ 5 & 6 & 7 \\ 8 & 9 & 1 \end{pmatrix}, \quad B = \begin{pmatrix} 2 & 3 & 4 \\ 5 & 4 & 3 \\ 1 & 2 & 1 \end{pmatrix}.$$

(a) Compute the adjoint matrix of each.
(b) Compute the inverse matrix of each.

Solution. (a) Recall that $adj(A)_{ij} = (-1)^{i+j} M_{ji}$. Using that formula, we obtain the following results

$$adj(A) = \begin{pmatrix} -57 & 33 & -3 \\ 51 & -30 & 6 \\ -3 & 6 & -3 \end{pmatrix},$$

$$adj(B) = \begin{pmatrix} -2 & 5 & -7 \\ -2 & -2 & 14 \\ 6 & -1 & -7 \end{pmatrix}.$$

(b) Since $A^{-1} = adj(A)/det(A)$, then all we need is to compute the determinants.

$$det(A) = -57 det \begin{pmatrix} -30 & 6 \\ 6 & -3 \end{pmatrix} - 33 det \begin{pmatrix} 51 & 6 \\ -3 & -3 \end{pmatrix} - 3 det \begin{pmatrix} 51 & -30 \\ -3 & 6 \end{pmatrix} = 27,$$
$$det(B) = -2 det \begin{pmatrix} -2 & 14 \\ -1 & -7 \end{pmatrix} - 5 det \begin{pmatrix} -2 & 14 \\ 6 & -7 \end{pmatrix} - 7 det \begin{pmatrix} -2 & -2 \\ 6 & -1 \end{pmatrix} = 14.$$

Hence

$$A^{-1} = \frac{1}{27} \begin{pmatrix} -57 & 33 & -3 \\ 51 & -30 & 6 \\ -3 & 6 & -3 \end{pmatrix}, \quad B^{-1} = \frac{1}{14} \begin{pmatrix} -2 & 5 & -7 \\ -2 & -2 & 14 \\ 6 & -1 & -7 \end{pmatrix}.$$

Exercise 8.12

Use Cramer's method to solve the following system of linear equations:

$$x_1 + 4x_2 + 2x_3 = 8,$$
$$2x_1 - 3x_2 + x_3 = 12,$$
$$6x_1 + x_2 - 8x_3 = -29.$$

Solution. Using Cramer's rule, $x_i = det(A_i)/det(A)$, where A_i is identical to A except the ith column, which is equal to b. In our case,

$$A = \begin{pmatrix} 1 & 4 & 2 \\ 2 & -3 & 1 \\ 6 & 1 & -8 \end{pmatrix}, \quad b = \begin{pmatrix} 8 \\ 12 \\ -29 \end{pmatrix}.$$

Then

$$det(A) = 1 \cdot (3 \cdot 8 - 1) - 4 \cdot (-2 \cdot 8 - 6) + 2 \cdot (2 - (-3) \cdot 6) = 151.$$

This leads to:

$$det(A_1) = det \begin{pmatrix} 8 & 4 & 2 \\ 12 & -3 & 1 \\ -29 & 1 & -8 \end{pmatrix} = 302 \implies x_1 = 302/151,$$

$$det(A_2) = det \begin{pmatrix} 1 & 8 & 2 \\ 2 & 12 & 1 \\ 6 & -29 & -8 \end{pmatrix} = -151 \implies x_2 = -151/151,$$

$$det(A_3) = det \begin{pmatrix} 1 & 4 & 8 \\ 2 & -3 & 12 \\ 6 & 1 & -29 \end{pmatrix} = 755 \implies x_3 = 755/151.$$

Altogether,

$$\begin{pmatrix} x_1 \\ x_2 \\ x_3 \end{pmatrix} = \begin{pmatrix} 2 \\ -1 \\ 5 \end{pmatrix}.$$

13.9   Chapter 9

Exercise 9.1

(a) Let A, B be two square matrices. Show that both matrices AB and BA have the same eigenvalues.
(b) Show that the matrix $A \in \mathbf{R}^{n \times n}$ has the same characteristic polynomial as A^t.
(c) Show that the set of vectors $v \in R^n$ which satisfy $Av = \lambda v$ for a given $\lambda \in R$, form a linear sub-space of R^n.

(d) Let $A = \left(\begin{smallmatrix} 3 & -4 \\ 2 & -6 \end{smallmatrix}\right)$. Find all eigenvalues and corresponding eigenvectors of A.

Solution. (a) If v is an eigenvector of AB corresponding to eigenvalue λ, then it satisfies $ABv = \lambda v$. Multiplying both sides by B from the left and we obtain $BA(Bv) = \lambda Bv$. Thus, BA has the same eigenvalue (but not the same corresponding eigenvector).

(b) We now form previous chapters that $det(A) = det(A^t)$. The same applies for characteristic polynomials, i.e., $det(\lambda I - A) = det((\lambda I - A)^t) = det(\lambda I - A^t)$. Thus, the characteristic polynomial of A and A^t are identical.

(c) Denote $V_\lambda = \{v \in R^n | Av = \lambda v\}$. We will show closure with respect to addition and multiplication by a scalar. For every $v \in V_\lambda$ and $\alpha \in R$ we have:

$$A(\alpha v) = \alpha(Av) = \alpha \lambda v = \lambda \alpha v \implies \alpha v \in V_\lambda.$$

Now, for every $v_1, v_2 \in V_\lambda$ we have:

$$A(v_1 + v_2) = Av_1 + Av_2 = \lambda v_1 + \lambda v_2 = \lambda(v_1 + v_2) \implies v_1 + v_2 \in V_\lambda.$$

(d)

$$det(\lambda I - A) = det \left(\begin{smallmatrix} \lambda-3 & 4 \\ -2 & \lambda+6 \end{smallmatrix}\right)$$
$$= (\lambda - 3)(\lambda + 6) + 8 = \lambda^2 + 3\lambda - 10 = (\lambda + 5)(\lambda - 2),$$

which implies that $\lambda_1 = -5$ and $\lambda_2 = 2$. Next we compute the corresponding eigenvectors.

$$\begin{pmatrix} 3 & -4 \\ 2 & -6 \end{pmatrix} \begin{pmatrix} x \\ y \end{pmatrix} = \begin{pmatrix} -5x \\ -5y \end{pmatrix} \iff \begin{cases} 3x - 4y & = -5x \\ 2x - 6y & = -5y \end{cases} \iff y = 2x,$$

and we can choose $v_1 = (1, 2)$. Similarly, for λ_2:

$$\begin{pmatrix} 3 & -4 \\ 2 & -6 \end{pmatrix} \begin{pmatrix} 2x \\ 2y \end{pmatrix} = \begin{pmatrix} -5x \\ -5y \end{pmatrix} \iff \begin{cases} 3x - 4y & = 2x \\ 2x - 6y & = 2y \end{cases} \iff 4y = x,$$

and we can choose $v_2 = (4, 1)$.

Exercise 9.2

Let:

$$A = \begin{pmatrix} 1 & 4 & 3 \\ 0 & 3 & 1 \\ 0 & 2 & -1 \end{pmatrix}.$$

(a) What is the characteristic polynomial of A?
(b) Prove that 1 is an eigenvalue of A.
(c) Are there any more eigenvalues? If so, compute them.

Solution. (a)

$$f_A(\lambda) = det \begin{pmatrix} \lambda-1 & -4 & -3 \\ 0 & \lambda-3 & -1 \\ 0 & -2 & \lambda+1 \end{pmatrix} = (\lambda - 1)(\lambda^2 - 2\lambda - 5).$$

(b) If we plug the value 1 in the above characteristic polynomial, it will vanish. Therefore, 1 is an eigenvalue.

(c) Solving $\lambda^2 - 2\lambda - 5 = 0$ yields $\lambda_{1,2} = 1 \pm \sqrt{6}$, which are eigenvalues as well.

Exercise 9.3

Prove that if the sum of the algebraic multiplicities of eigenvalues equals n (the dimension of the matrix), then the coefficient of x^{n-1} in the characteristic polynomial is minus the sum of the eigenvalues, where each is multiplied by its algebraic multiplicity.

Solution. Assuming k different eigenvalues, denote by m_i the algebraic multiplicity of the ith eigenvalue λ_i. The characteristic polynomial is therefore:

$$f_A(\lambda) = (\lambda - \lambda_1)^{m_1} + \cdots + (\lambda - \lambda_k)^{m_k}.$$

A simpler way to write our polynomial is by denoting $\lambda_1, \ldots, \lambda_n$ all n eigenvalues, including all multiplicities. We therefore have:

$$f_A(\lambda) = \prod_{i=1}^{n} (\lambda - \lambda_i).$$

It can be shown that the coefficient of λ^{n-1} is $a_{n-1} = -\sum_{i=1}^{n} \lambda_i$. Let us prove that by induction: for a 1-degree polynomial, it is a free coefficient, whose sum is minus the only eigenvalue. For a 2-degree polynomial,

$$f_A(\lambda) = (\lambda - \lambda_1)(\lambda - \lambda_2) = \lambda^2 - (\lambda_1 + \lambda_2)\lambda + \lambda_1\lambda_2,$$

and our claim therefore holds. Suppose now that for an $(n-1)$-degree polynomial, $a_{n-2} = -\sum_{i=1}^{n-1} \lambda_i$. Developing the nth degree polynomial:

$$f_A(\lambda) = \prod_{i=1}^{n}(\lambda - \lambda_i) = (\lambda - \lambda_n)\prod_{i=1}^{n-1}(\lambda - \lambda_i)$$

$$= (\lambda - \lambda_n)\left(\lambda^{n-1} - \left(\sum_{i=1}^{n-1}\lambda_i\right)\lambda^{n-1} + r^{(n-3)}\right)$$

$$= \lambda^n - \lambda_n\lambda^{n-1} - \left(\sum_{i=1}^{n-1}\lambda_i\right)\lambda^{n-1} + r^{n-2}$$

$$= \lambda^n - \left(\sum_{i=1}^{n}\lambda_i\right)\lambda^{n-1} + r^{n-2},$$

where $r^{(n-1)}$ and $r^{(n-2)}$ are $(n-1)$, $(n-2)$-degree polynomials, respectively. We therefore showed that $a_{n-1} = -\sum_{i=1}^{n}\lambda_i$ for every n. Back to algebraic multiplicities, it can be seen that $a_{n-1} = -\sum_{i=1}^{k} m_i\lambda_i$.

Exercise 9.4

Let A be the following matrix:

$$A = \begin{pmatrix} 2 & 1 & 0 & 0 \\ 0 & 2 & 0 & 0 \\ 0 & 0 & 1 & 1 \\ 0 & 0 & -2 & 4 \end{pmatrix}.$$

(a) Compute $det(A)$.
(b) Find the eigenvalues of A.
(c) Find the algebraic and geometric multiplicities of each of the above eigenvalues.
(d) Show that the claim proved in Exercise 9.3 holds for matrix A.

Solution. (a)

$$det(A) = 2 \cdot 2 \cdot (1 \cdot 4 + 1 \cdot 2) = 24.$$

(b)

$$f_A(\lambda) = det \begin{pmatrix} \lambda-2 & -1 & 0 & 0 \\ 0 & \lambda-2 & 0 & 0 \\ 0 & 0 & \lambda-1 & -1 \\ 0 & 0 & 2 & \lambda-4 \end{pmatrix}$$

$$= (\lambda - 2)^2((\lambda - 1)(\lambda - 4) + 2) = (\lambda - 2)^3(\lambda - 3),$$

and therefore the eigenvalues are $\lambda_1 = 3, \lambda_2 = 2$.

(b) The algebraic multiplicity of λ_1 is 1, and therefore its geometric multiplicity is 1 as well (recall that algebraic multiplicity is not smaller than the geometric multiplicity).

(c) The algebraic multiplicity of λ_2 is 3. Let us find its geometric multiplicity: for $Av = 2v$, we get the following set of equations:

$$2x_1 + x_2 = 2x_1,$$
$$2x_2 = 2x_2,$$
$$x_3 + x_4 = 2x_3,$$
$$-2x_3 + 4x_4 = 2x_4,$$

which yields $x_2 = 0$, $x_3 = x_4$. Thus, the eigenvector is of the form $(x_1, 0, x_3, x_3)$ where $x_1, x_3 \neq 0$, and therefore the geometric multiplicity of λ_2 is 2.

(d) Examining the characteristic polynomial from (b), we can see that the coefficient of λ^3 is -9. On the other hand, we can see that $-(1 \cdot \lambda_1 + 3 \cdot \lambda_2) = -9$.

Exercise 9.5

Let $A \in R^{2 \times 2}$ be a symmetric matrix, i.e., the $(1, 2)$ and $(2, 1)$ entries are identical (a formal definition will be given in the next chapter).

(a) Prove the existence of the eigenvalues of A.
(b) Find an expression for the eigenvalues as a function of the matrix entries $\begin{pmatrix} a & b \\ b & d \end{pmatrix}$.

(c) Show that the two eigenvalues are different if the main diagonal entries are different.

Solution. (a) Since A is symmetric, it can be written as $\left(\begin{smallmatrix} a & b \\ b & d \end{smallmatrix}\right)$. Computing the characteristic polynomial, we get

$$det \left(\begin{smallmatrix} \lambda-a & -b \\ -b & \lambda-d \end{smallmatrix}\right) = (\lambda - a)(\lambda - d) - b^2 = \lambda^2 - \lambda(a + d) + ad - b^2.$$

This polynomial has real number roots only if its discriminant is equal or larger than 0, namely

$$(a + d)^2 - 4(ad - b^2) \geq 0.$$

But then

$$
\begin{aligned}
(a + d)^2 - 4(ad - b^2) &= a^2 + 2ad + d^2 - 4ad + 4b^2 \\
&= a^2 - 2ad + d^2 + 4b^2 = (a - d)^2 + 4b^2 \geq 0,
\end{aligned}
$$

which in any case is non-negative, and therefore the eigenvalues exist.

(b) Based on (a), finding the roots is immediate

$$\lambda_1 = \frac{a + d + \sqrt{(a - d)^2 + 4b^2}}{2},$$

$$\lambda_2 = \frac{a + d - \sqrt{(a - d)^2 + 4b^2}}{2}.$$

(c) We can see that if the entries a and d are different, then the discriminant is always positive, which leads to two different roots, or eigenvalues.

Exercise 9.6

Let

$$A = \begin{pmatrix} 1 & -3 & 3 \\ 3 & -5 & 3 \\ 6 & -6 & 4 \end{pmatrix}.$$

(a) Show that the eigenvalues are 4 and -2.
(b) Show that the eigenspace that corresponds to eigenvalue 4 is of dimension 1. Find the right and left eigenvectors whose scalar product is 1.
(c) Show that the eigenspace that corresponds to eigenvalue -2 is of dimension 2. Find a basis for the right eigenvectors. *Hint:* show that

$x_1 = (1, 1, 0)$ and $x_2 = (1, 0, -1)$ span the eigenspace that corresponds to eigenvalue -2.

(d) Show that the right eigenvectors of eigenvalue 4 do not belong to the sub-space spanned by the basis found previously.

(e) Show that A is diagonalizable.

Solution. (a) We will show that with these values, the characteristic polynomial vanishes

$$\det \begin{pmatrix} \lambda-1 & 3 & -3 \\ -3 & \lambda+5 & -3 \\ -6 & 6 & \lambda-4 \end{pmatrix} = (\lambda + 2)(\lambda^2 - 2\lambda - 8) = (\lambda + 2)^2(\lambda - 4).$$

(b) Note that the algebraic multiplicity of 4 is 1. Since the geometric multiplicity cannot be larger, we find that the geometric multiplicity of 4 is 1 as well. Thus, the dimension of the eigenspace of 4 is 1. Let us find the right and left eigenvectors

$$\begin{pmatrix} 1 & -3 & 3 \\ 3 & -5 & 3 \\ 6 & -6 & 4 \end{pmatrix} \begin{pmatrix} x \\ y \\ z \end{pmatrix} = \begin{pmatrix} 4x \\ 4y \\ 4z \end{pmatrix}.$$

The third equation in the system can be written as $6x - 6y = 0$ which implies that $x = y$. Plugging that in the first equation, we arrive at $z = 2x$. We can therefore choose $u = (1/2, 1/2, 1)^t$ as the right eigenvector. As for the left eigenvector, the system of equations becomes

$$\begin{pmatrix} x & y & z \end{pmatrix} \begin{pmatrix} 1 & -3 & 3 \\ 3 & -5 & 3 \\ 6 & -6 & 4 \end{pmatrix} = \begin{pmatrix} 4x & 4y & 4z \end{pmatrix}.$$

The solution has the form of $(-y, y, -y)^t$ so we can choose $v = (-1, 1, -1)^t$ as the left eigenvector. The inner product is $u^t v = -1$. To meet the question's requirements, we replace v with $-v$, the latter remains a left eigenvector of value 4, with the inner product $u^t(-v)$ now equals 1.

(c) We will find the right eigenspace of A, and show that it is spanned by two vectors. The corresponding system of equations is

$$\begin{pmatrix} 1 & -3 & 3 \\ 3 & -5 & 3 \\ 6 & -6 & 4 \end{pmatrix} \begin{pmatrix} x \\ y \\ z \end{pmatrix} = \begin{pmatrix} -2x \\ -2y \\ -2z \end{pmatrix},$$

which is equivalent to $x - y + z = 0$. Two vectors that satisfy this equation are $(1, 1, 0)$ and $(0, 1, 1)$, which are linearly independent, and span a two-dimensional subspace of R^3. The geometric multiplicity (the eigenspace dimension) therefore equals 2.

(d) We will show that any right eigenvector of eigenvalue 4 cannot be written as a linear combination of $(1, 1, 0)$ and $(0, 1, 1)$. We need to find coefficient a, b such that

$$\begin{pmatrix} 1/2 \\ 1/2 \\ 1 \end{pmatrix} = a \begin{pmatrix} 1 \\ 1 \\ 0 \end{pmatrix} + b \begin{pmatrix} 0 \\ 1 \\ 1 \end{pmatrix}.$$

One can easily see that no such solution exists, and therefore $(1/2, 1/2, 1)$ is not spanned by the above two vectors.

(e) The sum of the geometric multiplicities equals the matrix dimension, implying that the matrix is diagonalizable.

Exercise 9.7

Suppose that matrix A has the following spectral representation:

$$A = \sum_{i=1}^{n} \lambda_i E_i,$$

where λ_i is the eigenvalue, $E_i = x_i y_i^t$, x_i and y_i are the right and left eigenvectors, respectively.

(a) Prove that $\sum_{i=1}^{n} E_i = I$.
(b) Prove that for any x, which is not an eigenvalue of A, it holds that

$$(xI - A)^{-1} = \sum_{i=1}^{n} \frac{1}{x - \lambda_i} E_i.$$

Solution. (a) We will show that for any vector x,

$$\sum_{i=1}^{n} E_i x = Ix = x,$$

which will imply $\sum_{i=1}^{n} E_i x = I$. Given coefficients $\alpha_1, \ldots, \alpha_n$ such that $x = \sum_{i=1}^{n} \alpha_i x_i$ (recall that $x_1, \ldots, x_n$ is a basis). Now,

$$\sum_{i=1}^{n} E_i x = \sum_{i=1}^{n} x_i y_i^t \sum_{j=1}^{n} \alpha_j x_j = \sum_{i=1}^{n} x_i \sum_{j=1}^{n} \alpha_j y_i^t x_j = \sum_{i=1}^{n} \alpha_i x_i = x,$$

where the third equality (from the left) follows from $y_i^t x_i = 1$ and $y_i^t x_j = 0$ for $i \neq j$.

(b) We will show that $\left(\sum_{i=1}^{n}\frac{1}{x-\lambda_i}E_i\right)(xI - A) = I$

$$\left(\sum_{i=1}^{n}\frac{1}{x-\lambda_i}E_i\right)(xI - A) = \left(\sum_{i=1}^{n}\frac{1}{x-\lambda_i}E_i\right)\left(xI - \sum_{i=1}^{n}\lambda_i E_i\right)$$

$$= \sum_{i=1}^{n}\frac{1}{x-\lambda_i}E_i xI - \sum_{i=1}^{n}\left(\frac{1}{x-\lambda_i}E_i\sum_{j=1}^{n}\lambda_j E_j\right)$$

$$= \sum_{i=1}^{n}\frac{x}{x-\lambda_i}E_i - \sum_{i=1}^{n}\frac{\lambda_i}{x-\lambda_i}E_i^2$$

$$= \sum_{i=1}^{n}\frac{x}{x-\lambda_i}E_i - \sum_{i=1}^{n}\frac{\lambda_i}{x-\lambda_i}E_i$$

$$= \sum_{i=1}^{n}\frac{x-\lambda_i}{x-\lambda_i}E_i = \sum_{i=1}^{n}E_i = I.$$

The third equality follows from $E_i E_j = x_i y_i^t x_j y_j^t = 0$, $\forall i \neq j$, and the fourth equality follows from $E_i^2 = x_i y_i^t x_i y_i^t = x_i y_i^t$.

13.10   Chapter 10

Exercise 10.1

Let

$$A = \begin{pmatrix} 2 & 1 \\ 1 & 3 \end{pmatrix}.$$

(a) Find the eigenvalues of A.

(b) What is the value of the following objective? $\max_{x:\|x\|=1} x^t A x$.

(c) What is the value of the following objective? $\min_{x:\|x\|=1} x^t A x$.

(d) For each of the above problems, (b) and (c), find the vectors (x_1, x_2) that optimize the objectives.

(e) What is the spectral representation of A?

(f) Prove that if any matrix B is symmetric, then so is B^k for any integer k (including negative).

(g) Repeat (a)–(e) on A^{-1} and A^2 (hint: use any knowledge on A instead of computing the matrices A^{-1} and A^2).

Solution. (a) In Exercise 9.5, we developed closed form expressions for the eigenvalues of a general symmetric 2×2 matrix. Using this result here, we get

$$\lambda_1 = \frac{a + d + \sqrt{(a - d)^2 + 4b^2}}{2} = \frac{5 + \sqrt{5}}{2},$$

$$\lambda_2 = \frac{a + d - \sqrt{(a - d)^2 + 4b^2}}{2} = \frac{5 - \sqrt{5}}{2}.$$

(b) The maximal value is the largest eigenvalue of A, in our case $\lambda_1 = \frac{5+\sqrt{5}}{2}$.

(c) The minimal value is the smallest eigenvalue of A, in our case $\lambda_2 = \frac{5-\sqrt{5}}{2}$.

(d) Denote by v_1, v_2 the eigenvectors which match the eigenvalues λ_1, λ_2. It holds that

$$\lambda_1 = \max_{x: \|x\|=1} x^t A \text{ and } \arg\max_{x: \|x\|=1} x^t A = \frac{v_1}{\|v_1\|}, \tag{10.1}$$

and

$$\lambda_2 = \min_{x: \|x\|=1} x^t A \text{ and } \arg\min_{x: \|x\|=1} x^t A = \frac{v_1}{\|v_1\|}. \tag{10.2}$$

Thus, all we need to do is to find these eigenvectors and normalize them

$$\begin{pmatrix} 2 & 1 \\ 1 & 3 \end{pmatrix} \begin{pmatrix} x \\ y \end{pmatrix} = \begin{pmatrix} \lambda_1 x \\ \lambda_1 y \end{pmatrix} \iff y = \frac{1 + \sqrt{5}}{2} x,$$

so we can select $v_1 = \begin{pmatrix} 1 \\ \frac{1+\sqrt{5}}{2} \end{pmatrix}$, and therefore $\frac{v_1}{\|v_1\|} = \frac{1}{1.9} \begin{pmatrix} 1 \\ \frac{1+\sqrt{5}}{2} \end{pmatrix}$. Similarly,

$$\begin{pmatrix} 2 & 1 \\ 1 & 3 \end{pmatrix} \begin{pmatrix} x \\ y \end{pmatrix} = \begin{pmatrix} \lambda_2 x \\ \lambda_2 y \end{pmatrix} \iff y = \frac{1 - \sqrt{5}}{2} x,$$

selecting $v_2 = \begin{pmatrix} 1 \\ \frac{1-\sqrt{5}}{2} \end{pmatrix}$, we arrive at $\frac{v_2}{\|v_2\|} = \frac{1}{1.17} \begin{pmatrix} 1 \\ \frac{1-\sqrt{5}}{2} \end{pmatrix}$.

(e) Recall that the spectral representation is defined as $\sum_{i=1}^n \lambda_i x_i y_i^t$, where x_i is the right eigenvalues that matches λ_i, while y_i is the left eigenvalues that matches λ_i. Since A is symmetric, the right and left eigenvectors coincide, and our spectral representation becomes:

$$\lambda_1 x_1 x_1^t + \lambda_2 x_2 x_2^t$$

$$= \frac{5+\sqrt{5}}{2} \frac{1}{1.9^2} \begin{pmatrix} 1 \\ \frac{1+\sqrt{5}}{2} \end{pmatrix} \begin{pmatrix} 1 & \frac{1+\sqrt{5}}{2} \end{pmatrix} + \frac{5-\sqrt{5}}{2} \frac{1}{1.17^2} \begin{pmatrix} 1 \\ \frac{1-\sqrt{5}}{2} \end{pmatrix} \begin{pmatrix} 1 & \frac{1-\sqrt{5}}{2} \end{pmatrix}$$

$$= \frac{5+\sqrt{5}}{2} \frac{1}{1.9^2} \begin{pmatrix} 1 & \frac{1+\sqrt{5}}{2} \\ \frac{1+\sqrt{5}}{2} & \frac{3+\sqrt{5}}{2} \end{pmatrix} + \frac{5-\sqrt{5}}{2} \frac{1}{1.17^2} \begin{pmatrix} 1 & \frac{1-\sqrt{5}}{2} \\ \frac{1-\sqrt{5}}{2} & \frac{3-\sqrt{5}}{2} \end{pmatrix}.$$

(f) We prove first through induction on positive k. It holds for $k = 1$ by definition: $A^t = A$. For $k = 2$

$$A^2 = AA = A^t A^t = (AA)^t = (A^2)^t.$$

Suppose that it is true for $k - 1$. We will prove that for k:

$$A^k = A^{k-1} A = (A^{k-1})^t A^t = (AA^{k-1})^t = (A^k)^t.$$

Now, for negative k, we will show that if A is symmetric then so is A^{-1}. This will imply that A^k is symmetric for $k < 0$, since A^{-k} is symmetric. It holds that:

$$AA^{-1} = I \implies (AA^{-1})^t = I^t = I,$$

and therefore

$$(AA^{-1})^t = AA^{-1} \implies (A^{-1})^t A^t = A^{-1}A.$$

Since A is symmetric, $(A^{-1})^t A = A^{-1}A$, which implies that $(A^{-1})^t = A^{-1}$.

(g) The eigenvalues of A^{-1} are $\frac{1}{\lambda_1}$ and $\frac{1}{\lambda_2}$ for the same eigenvectors x_i. Therefore,

$$\max_{x:\|x\|=1} x^t A^{-1} x = \frac{1}{\min_i \lambda_i} = \frac{2}{5-\sqrt{5}},$$

$$\min_{x:\|x\|=1} x^t A^{-1} x = \frac{1}{\max_i \lambda_i} = \frac{2}{5+\sqrt{5}},$$

and we have: $A^{-1} = \sum_{i=1}^{2} \frac{1}{\lambda_i} x_i x_i^t$. For A^2, the eigenvalues are λ_1^2 and λ_2^2 for the same eigenvectors x_i. Therefore,

$$\max_{x:\|x\|=1} x^t A^2 x = (\max_i \lambda_i)^2 = \left(\frac{5 + \sqrt{5}}{2}\right)^2,$$

$$\min_{x:\|x\|=1} x^t A^2 x = (\min_i \lambda_i)^2 = \left(\frac{5 - \sqrt{5}}{2}\right)^2.$$

Finally, $A^2 = \sum_{i=1}^{2} \lambda_i^2 x_i x_i^t$.

Exercise 10.2

Consider the following symmetric matrix:

$$V = \begin{pmatrix} 1 & 0 & \rho \\ 0 & 1 & 0 \\ \rho & 0 & 1 \end{pmatrix}.$$

(a) Show that the characteristic polynomial of V equals $P_V(x) = (x - 1)$ $[(x - 1)^2 - \rho^2]$.
(b) Show that there exists three eigenvalues: 1, $1 + \rho$, and $1 - \rho$.
(c) Conclude that the matrix is positive if and only if $-1 < \rho < 1$.

Solution. (a) The characteristic polynomial is

$$det\left(xI - \begin{pmatrix} 1 & 0 & \rho \\ 0 & 1 & 0 \\ \rho & 0 & 1 \end{pmatrix}\right) = det\begin{pmatrix} x-1 & 0 & -\rho \\ 0 & x-1 & 0 \\ -\rho & 0 & x-1 \end{pmatrix}$$

$$= (x - 1)det\begin{pmatrix} x-1 & 0 \\ 0 & x-1 \end{pmatrix} - \rho \cdot det\begin{pmatrix} 0 & x-1 \\ -\rho & 0 \end{pmatrix}$$

$$= (x - 1)^3 - \rho^2(x - 1) = (x - 1)[(x - 1)^2 - \rho^2].$$

(b) The roots of the above polynomial are: $x_1 = 1$, and $(x - 1)^2 = \rho^2$, which leads to: $x_{2,3} = 1 \pm \rho$, and these are the eigenvalues.

(c) The matrix is positive if and only if all its eigenvalues are positive, namely $1 + \rho > 0$ and $1 - \rho > 0$, which is equivalent to $-1 < \rho < 1$.

Exercise 10.3

Consider the following symmetric matrix:

$$V = \begin{pmatrix} 1 & \rho & \rho \\ \rho & 1 & 0 \\ \rho & 0 & 1 \end{pmatrix}.$$

(a) Let $x = (1, -\sqrt{2}/2, -\sqrt{2}/2)^t$. Which condition needs to be imposed on ρ in order to guarantee that $x^t V x > 0$.

(b) Repeat the previous item but now with $x = (1, \sqrt{2}/2, \sqrt{2}/2)^t$.

(c) What are the eigenvalues of V? Conclude that the two conditions you derived above are also sufficient for V being positive.

(d) Derive the corresponding eigenvectors and state the spectral representation of V.

Solution. (a) Deriving the expression for $x^t V x$

$$(1, -\sqrt{2}/2, -\sqrt{2}/2) \begin{pmatrix} 1 & \rho & \rho \\ \rho & 1 & 0 \\ \rho & 0 & 1 \end{pmatrix} \begin{pmatrix} 1 \\ -\sqrt{2}/2 \\ -\sqrt{2}/2 \end{pmatrix} = 1 - 2\sqrt{2}\rho + 1 = 2 - 2\sqrt{2}\rho,$$

it is positive if $\rho < 1/\sqrt{2}$.

(b) Similarly,

$$(1, \sqrt{2}/2, \sqrt{2}/2) \begin{pmatrix} 1 & \rho & \rho \\ \rho & 1 & 0 \\ \rho & 0 & 1 \end{pmatrix} \begin{pmatrix} 1 \\ \sqrt{2}/2 \\ \sqrt{2}/2 \end{pmatrix} = 1 + 2\sqrt{2}\rho + 1 = 2 + 2\sqrt{2}\rho,$$

which is positive if $\rho > -1/\sqrt{2}$.

(c) Lets compute the characteristics polynomial

$$\det(xI - V) = \det \begin{pmatrix} x-1 & -\rho & -\rho \\ -\rho & x-1 & 0 \\ -\rho & 0 & x-1 \end{pmatrix}$$

$$= -\rho \det \begin{pmatrix} -\rho & -\rho \\ x-1 & 0 \end{pmatrix} + (x-1)\det \begin{pmatrix} x-1 & -\rho \\ -\rho & x-1 \end{pmatrix}$$

$$= -\rho^2(x-1) + (x-1)[(x-1)^2 - \rho^2]$$

$$= (x-1)[(x-1)^2 - 2\rho^2].$$

The eigenvalues are the roots of that polynomial, which are $x_1 = 1$ and $x_{2,3} = 1 \pm \sqrt{2}\rho$. We now can see that the eigenvalues are positive if $-1/\sqrt{2} < \rho < 1/\sqrt{2}$.

(d) To find the eigenvectors that match the values $1 \pm \sqrt{2}\rho$, we need to solve the following system of equations

$$x_1 + \rho x_2 + \rho x_3 = (1 \pm \sqrt{2}\rho)x_1,$$
$$\rho x_1 + x_2 = (1 \pm \sqrt{2}\rho)x_2, \tag{10.3}$$
$$\rho x_1 + x_3 = (1 \pm \sqrt{2}\rho)x_3.$$

Note that if $\rho = 0$, then we get the identity matrix, and any vector is an eigenvector of eigenvalue 1. Suppose then that $\rho \neq 0$, then the above set of equations is equivalent to

$$\pm\sqrt{2}x_1 + x_2 + x_3 = 0,$$
$$x_1 \pm \sqrt{2}x_2 = 0,$$
$$x_1 \pm \sqrt{2}x_3 = 0.$$

Solving for $\sqrt{2}$, we obtain the vector $\begin{pmatrix} 1 \\ -\sqrt{2}/2 \\ -\sqrt{2}/2 \end{pmatrix}$, as in (a). Solving for $-\sqrt{2}$, we obtain the vector $\begin{pmatrix} 1 \\ \sqrt{2}/2 \\ \sqrt{2}/2 \end{pmatrix}$. As for the eigenvalue 1, we need to replace the right-hand side of the system (10.3) with $\begin{pmatrix} x_1 \\ x_2 \\ x_3 \end{pmatrix}$. This system of equations is equivalent to

$$x_2 + x_3 = 0,$$
$$x_1 = 0,$$
$$x_1 = 0,$$

whose obvious orthonormal solution is $\begin{pmatrix} 0 \\ -\sqrt{2}/2 \\ \sqrt{2}/2 \end{pmatrix}$. The spectral representation then becomes

$$\begin{pmatrix} 0 \\ -\sqrt{2}/2 \\ \sqrt{2}/2 \end{pmatrix} (0, -\sqrt{2}/2, \sqrt{2}/2) + (1 + \sqrt{2}\rho) \begin{pmatrix} 1 \\ -\sqrt{2}/2 \\ -\sqrt{2}/2 \end{pmatrix} (1, -\sqrt{2}/2, -\sqrt{2}/2)$$
$$+ (1 - \sqrt{2}\rho) \begin{pmatrix} 1 \\ \sqrt{2}/2 \\ \sqrt{2}/2 \end{pmatrix} (1, \sqrt{2}/2, \sqrt{2}/2)$$
$$= \begin{pmatrix} 0 & 0 & 0 \\ 0 & 1/2 & -1/2 \\ 0 & -1/2 & 1/2 \end{pmatrix} + (1 + \sqrt{2}\rho) \begin{pmatrix} 0 & -\sqrt{2}/2 & -\sqrt{2}/2 \\ -\sqrt{2}/2 & 1/2 & 1/2 \\ -\sqrt{2}/2 & 1/2 & 1/2 \end{pmatrix}$$
$$+ (1 - \sqrt{2}\rho) \begin{pmatrix} 0 & \sqrt{2}/2 & \sqrt{2}/2 \\ \sqrt{2}/2 & 1/2 & 1/2 \\ \sqrt{2}/2 & 1/2 & 1/2 \end{pmatrix}.$$

Exercise 10.4

A matrix A is skew-symmetric if $A_{ij} = -A_{ji}$ for any $i \neq j$.

(a) Show that if λ is an eigenvalue of A then it is zero.
(b) Does that imply that A is singular? Prove if so, otherwise find a counter-example.

Solution. (a) Since A is skew-symmetric, then $A^t = -A$. If λ is an eigenvalue then it holds that $Av = \lambda v$ for some vector v, and $v^t A v = \lambda \|v\|^2$. On the other hand, applying transposition on $v^t A v$ we get $v^t A v = v^t A^t v = -\lambda \|v\|^2$. Since $\|v\|^2 \neq 0$, we get $\lambda = -\lambda = 0$.

(b) This is wrong, for example $\left(\begin{smallmatrix} 1 & 2 \\ -2 & 1 \end{smallmatrix}\right)$ is skew-symmetric but invertible. Note that A here does not have eigenvalues.

Exercise 10.5

Let $A = \left(\begin{smallmatrix} 2 & -2 \\ -2 & 5 \end{smallmatrix}\right)$ be a symmetric matrix. The matrix represents linear transformation on the elementary basis.

(a) Find a matrix P whose columns form an orthonormal basis of R^2, such that the transformation matrix is diagonal.
(b) How can you find P^{-1} in that case, without calculating the inverse matrix?
(c) What is the spectral representation of A?

Solution. (a) We form a matrix P whose column are the normalized eigenvectors of A, and the transformation matrix is the diagonal with the eigenvalues in its main diagonal. We know that if a symmetric matrix has different values in its main diagonal, then it has different eigenvalues

$$\lambda_{1,2} = \frac{a + d \pm \sqrt{(a-d)^2 + 4b^2}}{2} = \frac{7 \pm \sqrt{25}}{2} \implies \lambda_1 = 6, \lambda_2 = 1.$$

Let's find the right eigenvectors for each of the above eigenvalues:

$$\begin{pmatrix} 2 & -2 \\ -2 & 5 \end{pmatrix} \begin{pmatrix} x_1 \\ x_2 \end{pmatrix} = \begin{pmatrix} 6x_1 \\ 6x_2 \end{pmatrix} \iff \begin{cases} 2x_1 - 2x_2 & = 6x_1 \\ -2x_1 + 5x_2 & = 6x_2 \end{cases} \iff x_2 = -2x_1,$$

we can choose $v_1 = (1, -2)$.

$$\begin{pmatrix} 2 & -2 \\ -2 & 5 \end{pmatrix} \begin{pmatrix} x_1 \\ x_2 \end{pmatrix} = \begin{pmatrix} x_1 \\ x_2 \end{pmatrix} \iff \begin{cases} 2x_1 - 2x_2 & = x_1 \\ -2x_1 + 5x_2 & = x_2 \end{cases} \iff x_1 = 2x_2$$

choose $v_2 = (2, 1)$. Finally, the orthonormal matrix P becomes:

$$P = \frac{1}{\sqrt{5}} \begin{pmatrix} 1 & 2 \\ -2 & 1 \end{pmatrix} \implies P^{-1}AP = \begin{pmatrix} \lambda_1 & 0 \\ 0 & \lambda_2 \end{pmatrix}.$$

(b) Since P is orthonormal then $P^{-1} = P^t$.

(c) For the normalized v_1, v_2 we have:

$$E_1 = v_1 v_1^t = \frac{1}{5} \begin{pmatrix} 1 & -2 \\ -2 & 4 \end{pmatrix},$$

$$E_2 = v_2 v_2^t = \frac{1}{5} \begin{pmatrix} 4 & 2 \\ 2 & 1 \end{pmatrix},$$

so that $A = \lambda_1 E_1 + \lambda_2 E_2$.

Exercise 10.6

Let $A = \begin{pmatrix} a+b & b \\ b & c+b \end{pmatrix}$ be a symmetric matrix with $a, b, c \geq 0$. Prove that A is a semi-positive matrix.

Solution. For $x \in R^2$

$$a^t A x = ax_1^2 + b(x_1^2 + 2x_1x_2 + x_2^2) + cx_2^2 = ax_1^2 + b(x_1 + x_2)^2 + cx_2^2,$$

which of course is not negative under the exercise's assumptions.

13.11 Chapter 12

Exercise 12.1

1. Show that if $x \in R^n$, obeys $x_i P_{ij} = x_j P_{ji}$, $1 \leq i, j \leq n$ (called the *detailed balance equations*, then it also solves the balance equations.
2. Give an example where the detailed balance equations do not have a solution.
3. Show that if and only if for any triple of nodes i, j and K, $1 \leq i, j, k \leq n$, $P_{ij} P_{jk} P_{ki} = P_{ik} P_{kj} P_{ji}$, the detailed balance equations have a unique solution.

Solution.

1. Summing up with respect to i, the condition $x_j P_{ji} = x_i P_{ij}$, we get (since $\Sigma_i P_{ji} = 1$), $x_j = \Sigma_i x_i P_{ij}$, which is a balance equation.

2. bla

3. Assume the condition holds. For some state i_0, fix without loss of generality, $x_{i_0} = 1$. Now try the solution $x_i = P_{i_0,i}/P_{i,i_0}$. We need to show that $x_i P_{ij} = x_j P_{ji}$. Given this guess, the left-hand side equals $\frac{P_{i_0,i}}{P_{i,i_0}} P_{ij}$, while the right-hand side equals $\frac{P_{i_0,j}}{P_{j,i_0}} P_{ji}$. This is easily seen to hold by the assumed assumption. For the converse, let $\pi \in R^n$ be a solution to the detailed balanced equations (which without loss of generality comes with the limit probabilities). Then for $i, j, k,\ 1 \leq i, j, k \leq n$,

$$\pi_i P_{ij} = \pi_j P_{ji},$$

$$\pi_j P_{jk} = \pi_k P_{kj},$$

and

$$\pi_k P_{ki} = \pi_i P_{ik}.$$

Multiplying all right-hand sides and then all left-hand sides, leads to the required condition.

Exercise 12.2

Let $(A, \overline{A})$ be a partition of the state space. Show that

$$\sum_{i \in A} x_i \sum_{j \in \overline{A}} P_{ij} = \sum_{i \in \overline{A}} x_i \sum_{j \in A} P_{ij},$$

for any partition if and only if x is a constant multiplier for π.

Solution. One direction of the proof is trivial: Take $A = \{i\}$ and $\overline{A} = \{1, 2, \ldots, i-1.i+1, \ldots, n\}$. For the converse, let $x \in R^n$ obey $x_j = \Sigma_i x_i P_{ij}$. Summing up across $j \in A$, we get

$$\sum_{j \in A} x_j \sum_i P_{ji} = \sum_{j \in A} \sum_i x_i P_{ij},$$

and

$$\sum_{j \in A} x_j \left(\sum_{i \in A} P_{ji} + \sum_{i \in \overline{A}} P_{ji} \right) = \sum_{j \in A} \left(\sum_{i \in A} x_i P_{ij} + \sum_{i \in \overline{A}} x_i P_{ij} \right).$$

The double summation with respect to A in both hand sides canceled each other. What is left is what was requested.

Exercise 12.3

For the matrix

$$S = \begin{pmatrix} 0 & 1 & 0 & 0 & 0 \\ 1 & 0 & 0 & 0 & 0 \\ 0 & 0 & 0 & 1 & 0 \\ 0 & 0 & 0 & 0 & 1 \\ 0 & 0 & 1 & 0 & 0 \end{pmatrix},$$

compute S^{150} and S^{1001} (*Hint*: find the periodicity).

Solution. We note two important things. First, $S^6 = S^{12} = S^{18} = \cdots = I$. That is, $S^{6n} = I$ (equal to the identity matrix) for any $n = 1, 2, \ldots$. Second, standard computation shows that $S^t S = I$. It follows from the first property that $S^{150} = S^{6 \cdot 25} = I$. Finally, note that 1002 is a product of 6, and therefore $S^{1002} = I$. On the other hand, $S^{1002} = S^{1001} S$, so that S^{1001} is the inverse of S. Because S^t is also the inverse of S, then $S^{1001} = S^t$.

Bibliography

[1] Ash, R., *Basic Probability Theory*, John Wiley & Sons, Inc., New York, 1970.

[2] Bertsekas, D.P., *Nonlinear Programming*, Athena Scientific, 2016.

[3] Boyd, B. and L. Vanderberghe, *Introduction to Linear Algebra: Vectors, Matrices and Least Squares*, Cambridge University Press, 2018.

[4] Chigansky, P., Personal communication, 2022.

[5] Hui, J. Machine learning — Singular value decomposition (SVD) and principal component analysis (PCA), https://jonathan-hui.medium.com/machine-learning-singular-value-decomposition-svd-principal-component-analysis-pca-1d45e885e491.

[6] Lavrov, M., https://faculty.math.illinois.edu/~mlavrov/docs/484-spring-2019/ch1lec5.pdf.

[7] Neter, J., W. Wasserman and G.A. Whitemore, *Applied Statistics*, 3rd edition, Allyn and Bacon, Inc, 1988.

[8] Lipschutz, S. and M. Lipson, *Linear Algebra*, 4th edition, Schaum's Outline Series, The McGraw-Hill Company, New York, 2009.

[9] Seneta, E., *Non-negative Matrices and Markov Chains*, Springer Series in Statistics, 2006.

[10] Vanderberghe, L., http://www.seas.ucla.edu/~vandenbe/133A/lectures/chol.pdf.

Index

www.ingramcontent.com/pod-product-compliance
Lightning Source LLC
Chambersburg PA
CBHW050553190326
41458CB00007B/2022